The Wild Within

The Wild Within

Histories of a Landmark British Zoo

 ANDREW FLACK

University of Virginia Press | Charlottesville and London

University of Virginia Press
© 2018 by the Rector and Visitors of the University of Virginia
All rights reserved
Printed in the United States of America on acid-free paper

First published 2018

ISBN 978-0-8139-4093-9 (cloth)
ISBN 978-0-8139-4095-3 (ebook)

9 8 7 6 5 4 3 2 1

Library of Congress Cataloging-in-Publication Data is available from the Library
of Congress

Cover photo: Detail of lion cage with crowds, c. 1905. (Roy Vaughan Collection, Bristol
Zoological Society Archive, courtesy of the Bristol Zoological Society)

For Jessica, Oscar, and Evelyn

I think I could turn and live with animals, they are so placid and self-contain'd
I stand and look at them long and long.
They bring me tokens of myself.
—Walt Whitman, "Song of Myself"

The image of a wild animal becomes the starting point of a daydream: a point
from which the daydreamer departs with his back turned.
—John Berger, "Why Look at Animals?"

Contents

Acknowledgments

I want to thank the Arts and Humanities Research Council, whose generous funding via a Collaborative Doctoral Award between 2010 and 2013 allowed me the time to delve deep into a wonderfully rich archive, stored in a dusty attic at Bristol Zoo. The Bristol Zoological Society, of course, allowed me privileged access, not only to their historical records, but also to past and present staff members and current records, too. In particular, I thank Jo Gipps, who set this project in motion, Bryan Carroll and Kristina Haldane for granting access to archives and other "precious things." Christoph Schwitzer and Simon Garrett were both instrumental, not only in this work, but also in my development as a scholar interested in animals and environments. Their deep knowledge of wild things and wild places enriched my understanding of the planet, and I thank them for that. In addition, I thank all those who took part in the interviews upon which some of this work is based. These people, in particular, have formed the human world of the Bristol Zoo for the past seventy years or so.

I also owe thanks to various archives and public bodies whose materials form the basis of this work and who have granted me the opportunity to reproduce some of it here. These include Bristol Museum and Art Gallery; the Zoological Society of London; the Natural History Museum, London; Bristol Archives; Andre Pattenden; Hylton Warner; Victoria Arrowsmith-Brown; National Museums of Wales.

Over the course of this project, I took part in a number of public engagement activities. I therefore thank the British Broadcasting Corporation, the Pervasive Media Studio (Bristol), ThoughtDen (Bristol), Play Nicely (Bristol), Bristol Museum and Art Gallery (Bonnie Griffin and Rhian Rowson, in particular), and Research Enterprise and Development at the University of Bristol.

Various aspects of this study have appeared in alternative forms and contexts in the following publications: "Science, 'Stars' and Sustenance: The Acquisition and Display of Animals at the Bristol Zoological Gardens, 1836–c. 1970," in *Wild Things: Nature and the Social Imagination,* edited by William Beinart, Simon Pooley, and Karen Middleton, 163–84 (Cambridge: White Horse Press, 2013); "In Sight, Insane: Animal Agency, Captivity and the Frozen Wilderness in the Late Twentieth Century," *Environment and History* 22, no. 4 (2016): 629–52; "Capturing the Beasts: Zoo Film and Interspecies Pasts," in *The Zoo and Screen Media: Images of Exhibition and Encounter,* edited by Michael Lawrence and Karen Lury, 23–42 (Basingstoke: Palgrave Macmillan, 2016).

I owe a huge debt to two scholars who supervised the doctoral research underpinning this work, and who have offered guidance to a scholar whose propensity to be distracted by new beasts on the block must have been infuriating. Peter Coates supplied rooibos tea, biscuits, direction, and wisdom in apparently limitless quantities. I could not have asked for a more inspiring mentor, and for that I will always be grateful. Tim Cole gave me the space to experiment with ideas at the margins of historical studies. The combination of these men's expertise has made me into the scholar that I am today. The school of Coates and Cole has much to answer for. I also thank Sarah Joy Maddeaux, who worked alongside me on the Arts and Humanities Research Council Collaborative Doctoral Award. Her meticulous attention to detail and superhuman organizational abilities are beyond words. I learned a huge amount from her. In addition to them, I want to thank colleagues and mentors at the University of Bristol: Brendan Smith for having faith in me at the start, and Robert Skinner, Kirsty Reid, Simon Potter, Hilary Carey, Ronald Hutton, and Josie McLellan for wonderful conversations and their guidance over the years. I also want to applaud my past students who have, I hope, learned to recognize the importance of animals past, and whose enthusiasm and curiosity challenged and inspired me in equal measure.

I thank Boyd Zenner at the University of Virginia Press for her patience, encouragement, and humor. I am also grateful to the two anonymous reviewers for their rigorous reading of this manuscript. Their thoughts have enriched this work immeasurably. I owe special thanks to Alice Would and John Reeks for (having been co-opted into) reading this manuscript in its entirety and for having offered some valuable guidance. I also, of course, need to thank an entire flock of scholars for their insights over the years, at the university, at conferences, workshops and seminars during which I harassed them about "my zoo." These include Helen Cowie, Marianna Dudley, Andrea Gaynor, Daniel Haines, Jeffrey Hyson, Karen Jones, Dolly and Finn Arne Jørgensen, Hilda Kean, Rob Lambert, Garry Marvin, Margery Masterson, Susan McHugh, Chris Pearson, Harriet Ritvo, Nigel Rothfels, Jenna, Gareth, Lisa Uddin, Dan Vandersommers, and Peter Yeandle. There are many other people with whom I have shared fascinating conversations over the course of this book's writing. I thank them all for listening and guiding.

I also want to thank friends and family for their support and their interest, even if they imagined I was spending my time conducting interviews with gorillas (maybe I will, soon). Fran, Niall, Ollie, Katie, Michael, Jenna, Gareth, Nick, John, Rachel, Dan, Sarah, Rich, and Leah: thank you. I thank my family—and my grandfather John S. Moore in particular—for having believed that I could do this, and Jessica's family for having taken me on. I understand what an undertaking that represents. To all of my friends and family: thank you for riding the rough waters of these zoo years. This book was born from the depths, and, indeed, this book is because of you.

Many of the readers of this book might not be aware that there is something about me that can't be seen. I see the world differently: the world is blurring, slowly but surely. My wonderful wife, Jessica, has been by my side almost from the beginning of this. I owe her an enormous debt. She has never let me imagine for a moment that I am any less capable for it, nor that there is anything that I should not be able to do. She has shown me that seeing differently does not equate to a lack of vision. She pushes me ahead, has my back, and gives me the courage to accept the difference that I embody. I will always be grateful for that.

Finally, I want to acknowledge my children: Oscar and Evelyn. In part, this book is a token of my thanks for their having shown me a new kind of "wild" life. But it's also more than that. This is a warning that the wild world all around might not be all around forever. Keep it.

Introduction

That very night in Max's room a forest grew
and grew—
and grew until his ceiling hung with vines
and the walls became the world all around . . .
and when he came to the place where the wild things are
they roared their terrible roars and gnashed their terrible teeth
and rolled their terrible eyes and showed their terrible claws
'till Max said "BE STILL!"
and tamed them with the magic trick
of staring into their yellow eyes without blinking once
and they were frightened and called him the most wild thing of all . . .
—Maurice Sendak, *Where the Wild Things Are*

Where are the wild things? Where are they heading? *What are they?* These are vexing questions indeed, not least in our era of unprecedented environmental transformation at the hands of the voracious human animal. For much of the history of our species we have imagined that wild beasts dwelled in the regions of the Earth that remained untarnished by human activity, where nature was supposedly pure and uncorrupted.[1] Today, in the first decades of the twenty-first century, we witness wild things pushed to the margins of a world emblazoned with the footprints left by people.

Across Europe, over-exploitation, urban expansion, and development continue to have devastating consequences for native species. Grassland butterfly populations declined by 50 percent between 1990 and 2011, while common and farmland bird numbers have fallen by 12 and 30 percent, respectively, during the same period.[2] Further afield, polar bears roam southward into areas more densely populated with humans than their favored habitats on the roof of the world, while orangutans find themselves adrift on sylvan islands forged through the clearance of forests for palm oil plantations in Southeast Asia. The natural world is not a realm set apart from the activities of people. We live in a world where human and nonhuman spaces and systems entwine.

It is also a world where human relationships with wildlife are shaped by a diversity of attitudes around what an animal is and, indeed, what it means to be human. Across many of the world's cultures there has been a long-held assumption that nature and culture, human and animal, are absolute states that exist in distinction from each other. However, there is in fact no such simple division. The term "nature" itself is a culturally constructed category, an imagined entity, whose meaning has shifted in a diversity of ways over time and across societies.[3] Moreover, since humans *are* animals, then the products of human ingenuity must be the fruits of nature. Where do we draw the line between what is of nature and what is not?

In Maurice Sendak's classic 1963 children's story *Where the Wild Things Are,* fearsome creatures live in a remote place, on the other side of vast seas and oceans. It takes the story's young protagonist Max an arduous journey of nearly a year to reach the place where the wild things are. Yet, the truth is that all along the wild things and the wild places are very close at hand. Max's bedroom is the very place where the forest grows, where the walls transform into the wild world all around. All along, wild nature and wild beasts lurk inside the imagination of a little boy who was sent to bed without his dinner. In Sendak's charming tale, wild nature shows its colors. It is simultaneously everything and nothing at all, far away and near, domestic and exotic, real and imagined, for an imaginary world springs to life within the confines of Max's familiar and very tangible bedroom walls.

Indeed, what we often call wild is a particular kind of nature; it is a culturally constructed mode of understanding the world, a figment of the imagination that denotes what is beyond us and what is within us and that is, often terrifyingly, beyond our control. And because wildness emerges from the mind, it can, in fact, be anything. Wild is nature red in tooth and claw. Wild is the impenetrable rainforest. Wild is the nature that we have confined in our own mansions of wood, iron, glass, plastic, and stone. Wild is the domestic dog turned fierce, biting, snarling, or urinating on the carpet. Wild is the unsettling presence of delinquent youths or the psychologically distressed. Wild is everywhere. But what all of these wild natures have in common is a sense of separation and a lack of human control. It is the unknown and the unfamiliar.[4] Yet the ideological currents coursing through our notions of wildness are far from exceptional.

A host of human physical and imaginative relationships with wildlife and the rest of the natural world more broadly across time and space is characterized by the kind of ambiguities inherent in this rather slippery idea of the "wild." Many of us happily, ravenously, sometimes tenderly, place a chicken's dismembered leg, wing, or breast onto a scorching grill, though we are repulsed by the thought of doing the same to the bloody leg of a cute and cuddly feline or a loyal canine companion, even if he has just finished urinating on the carpet. Some of us refuse to chow down on the flesh of cows, sheep, and pigs but will delightedly tuck into a fish dinner. Still more of us are disheartened by the unrelenting destruction of habitats all over the world, all the way from South American rainforests to Pacific coral reefs, and yet we might still be keen to hack away at the greenery in our gardens, mowing and pruning and digging, swinging a wrecking ball through the habitations of the mini and the micro. Rarely do we ask ourselves the provocative question: why, exactly, is a wild rainforest desirable but a wild garden something that needs to be tamed, domesticated, and rendered liv*able* if rather less liv*ing*? Why should we eat some animals and not others? So many human relationships with wildlife are defined by, and riddled with, these kinds of contradictions. Animals and environments, living physical entities, are considered according to a multitude of cultural factors, which do not sit comfortably together. Our interactions with animal others reflect the very same entanglement of the real and the imagined, the familiar and the strange, the domestic and the exotic that waits in the cozy childish warmth of Max's wild bedroom.[5]

Much of the confusion embedded within these attitudes and entanglements rests on an enduring belief at the heart of some cultures: that humans are somehow intrinsically different from every other living thing: that organic collective known to us, despite the astonishing diversity among the world's living things, as animal.[6] This "abyssal rupture" has significant roots in Aristotelean philosophy and Judeo-Christian theology through the designation of nonhuman life as existing only for the purposes of the masterful human being.[7] When Max encounters the wild things, he is shown to be both human *and* animal; there is, in the end, no distinction. Yet, historically, the articulation of a fundamental distinction has been important in justifying a range of attitudes and behaviors toward animals, including (especially) exploitation. Employing his "magic trick"—his humanity—in order to tame the beasts, Max comes to direct them as a master choreographs the lurching movements of his puppets. Yet, he is called "the most wild thing of all" as he plays with them in their wild world. Through his interactions with the beasts, the barrier that has, historically speaking, been constructed to mark out the difference between human and animal is smashed to pieces. This distinction, and its fragility, sit at the very heart of a multitude of human-animal relationships in modernity. So powerful is it that trouble unfolds when the rupture closes and the borderlands dissolve.[8] Moments in which one of our own is devoured, or when our populations are threatened by nonhuman life-forms, even on the most microscopic of levels—such as the ongoing Ebola and Zika virus crises—profoundly shake the hubris of human separation from, and superiority over, the rest of the natural world. They rattle beliefs at the core of our being.[9] In less violent contexts, some animals, chiefly primates, look remarkably and often disturbingly like us in body and behavior. At the same time some humans, such as those who appeared on stage and, more recently, on screen in what are still often referred to as "freak shows," and those whose criminal behaviors have been considered so repulsive as to render them subhuman, have become residents of a liminal beastly realm.[10] This puzzling perception of sameness and difference, this closing of the abyssal rupture, underpins so many of our relationships with wildlife, from the ways in which we are able to empathize with some species more than others, to the moral effortlessness with which we can kill some of them for our own convenience. The combination of closeness and distance enables us to see animals as both objects for our own use and as individual subjects of lives that are so much like our own. The flexibility inherent in the abyssal rupture is unendingly useful to us because it permits the existence of diverse attitudes toward the nonhuman world.

Human relationships with animals are also vastly unbalanced. People have historically—and increasingly—wielded an inordinate degree of power over nonhuman lives and nonhuman representations. Yet, Max's encounters with the wild things are not merely about his mastery of them. Far from mute, the wild things look back and speak. They roar, roll their eyes, show their "terrible claws," and gnash their teeth in a nod to nature's capacity to dramatically influence human lives. Moreover, even once Max has (inevitably?) tamed them they continue to converse with him. Indeed, it is *they* who name him the "most wild thing of all," reflecting his true self right back at him. As Max's interactions with the wild things suggest, human-animal relationships are often about interaction and negotiation; humans and other animals converse in all sorts of ways, and it is frequently through these exchanges that the world—as complex as it is—is made. Such relations are a symptom of the connectedness of nature and culture, of human and animal and the artificiality of the abyssal rupture that seems to divide "our" world from the wild world all around.

In light of these complexities, the character of animal relationships with us—a species now changing the very geology of the Earth—has increasingly become a source of considerable interest and, indeed, unease.[11] Once a focus of amusement, scholarship relating to these interactions, inspired by the emerging public distaste for animal and environmental exploitation in the second half of the twentieth century, has grown exponentially and spans across disciplines reflecting the ubiquity of animals in lives past and present.[12] Of all the groups whose voices were acknowledged through the post-1960s social justice movements, and the democratization of academic enquiry, nonhuman animals might just be considered to have led the most marginalized of lives. Such a long-standing marginalization reflects a deep-rooted speciesism, a powerful prejudice comparable with those manifested in sexism, racism, ableism, and, indeed, all the "isms" of human social life, at the heart of Western cultures, in particular. Philosophers such as Richard Ryder and Peter Singer identified an innate imbalance of power in animal engagements with people (especially in the Western world), not least in the context of the animal-industrial complex through which animals are transformed into consumer goods on a colossal scale.[13] The recognition of the place of animals in human cultures across the span of space and time, as more than mere passers-by, has inspired their emergence from the metaphorical undergrowth of the academy in ever-increasing numbers.[14]

As we have already noted, however, nature is as much a figment of the imagination as it is physical entity. Consequently, animals as they actually are have sometimes been at risk of occlusion within the shadows they cast on human cultures. They really are, as Claude Lévi-Strauss famously remarked, "good to think [with]," and this renders their images infinitely malleable.[15] Nineteenth- and twentieth-century animal bodies were seen to be reflective of the concerns of Englishmen, for instance, from their ideas about hierarchies of life to imperial worldviews; snakes, for example, were construed as emblems of a certain kind of exoticism in the British Empire. Drenched with meaning ripe for extraction, nonhuman beings of all kinds have served an array of purposes in the past and, indeed, have come to serve scholars, too, as ciphers for an astonishing array of cultural idioms, ideas, and ideals.[16]

Interesting and important, certainly, but this attitude toward animals is problematic nonetheless because to overemphasize the shape-shifting nature of animals' reflections in the turbulent waters of human cultures is to risk obscuring the lived realities of rich animal lives. As entities of flesh and blood that stink, slurp, and shriek in embodied and emotional lives that are independent of human depictions, animals are much more than their cultural representations might have us believe.[17] As thinking and feeling beasts ourselves, we have little option but to consider animals biocentrically as richly emotional creatures in their own right. As Marc Bekoff powerfully observes, "viewing animals objectively does not work."[18] Comedian Chris Rock sardonically remarked on a tiger attack at a Las Vegas circus in 2003. His comments neatly capture how easy it remains for us to lose sight of the realities of animal being: "Everybody's mad at the tiger. 'Oh, the tiger went crazy.' No, he didn't. That tiger went *tiger*," he observed.[19] Animals on the largest and smallest of scales are physical presences, and their behaviors have been extremely influential in past events. They were instrumental in the European conquest of the New World, and they were major players in the success of frontiersmen as they opened up and began to manipulate the vast wildernesses of the American West.[20] Even the human-animal relationships we foster in our own domestic lives are forged through an elaborate interspecies dance. We and other animals are, as Donna Haraway so beautifully elucidates, bound together in an unending process of "becoming-with." Animals are capable of molding human attitudes and actions, and indeed, the wild things in Max's bedroom/mind certainly influence him in his play.[21] Animals ought, then, to be considered both imaginary, totemic beings, *and* physical influential presences, with lives of their very own.[22]

All of this brings us back to Max's bedroom but also to the gates of a historic institution rooted in space and time. Max's bedroom *is,* in fact, very much like the zoo. In the physical space demarcated by the four walls of the bedroom, a world of human-animal and human-environment relationships was simultaneously writ small *and* writ large—a "glocal" (from "global" and "local") space. His imagination formed a world inhabited by creatures who were both the same and different, who controlled and who were controlled, who were imaginary but who sprang to life in a very tangible physical space. There, relationships were performed. Such a magnificent multispecies affair characterizes the zoo, too. Of neither human nor nonhuman making, the space is a hybrid of nature and culture, of the so-called wild and the tame, of the real and the imagined, of life, death, and everything in between, just like the fabric of the world itself. In captivity, animals are imagined and manipulated, but they also look back. They are imaginary and physical presences in a cocreated domain: a "hybrid geography."[23]

Zoos and their forerunners have a long history.[24] Egyptian rulers established collections of exotic beasts circa 2,500 BCE as did the kings of Ur some four thousand years ago. Many centuries later Charlemagne, Frederick II, Henry I, and Louis XIV all maintained their own royal menageries, which often served as reflections of personal and stately power and prestige, and of diplomatic relations among rulers and states.[25] The modern zoo, a place traditionally defined by the tensions between science and amusement, however, emerged in Europe in the late eighteenth and early nineteenth centuries. The first is usually recognized as the imperial menagerie at Tiergarten Schönbrunn, Vienna (est. 1752). The Ménagerie du Jardin des Plantes in Paris (est. 1793) and the gardens of the Zoological Society of London in Regent's Park (opened in 1828) emerged in the years that followed, each with a prevailing scientific/experimental ethos. Such institutions arose during a historical moment characterized by the expansion of global connection and circulation. Like the enormous array of peoples, goods, and ideas that flowed from place to place within the globalizing systems at the core of the age of empires, the animals, many of which were icons of imperial mastery, that zoos were designed to exhibit were organic objects of desire that were increasingly ubiquitous within a vast global marketplace.[26] Indeed, the world itself was transforming into one laden with curious, marvelous, and accessible objects to be acquired and laid bare. The century's preoccupation with the fruits of nature, when the urban middle classes in Britain, and beyond, developed an almost insatiable desire to collect and organize the world's natural bounty, reflects both increasing interest and increasing availability in the globalizing marketplace.[27]

As well as this, in the early decades of the nineteenth century both recreational and scientific activities were growing in quantity and variety. Consequently, the attention of the—mostly—middle classes had turned toward the desire to provide display and experimentation spaces for the scientifically inclined so that the objects flowing from the exotic world into the domestic arena could be ordered and understood. At the same time middle-class reformers were occupied with the provision of enlightening recreational activities for the urban masses—"rational recreation"—whose penchant for entertainments such as those relating to the consumption of alcohol and the decadent pursuit of physical gratification were thought to have degenerative impacts on both body and soul. Educational recreational and increasingly formalized and codified sporting activities were supposed to widen the laborer's mental world, compensating for the repetitive, unfulfilling nature of much industrial work and providing an alternative to hedonism. Such activities also taught moral and technical lessons, encouraging the industriousness and innovation that were perceived to be vital to moral well-being and the health of the nation. This was an attempt to build a nation from the bottom up, part of which was through exposure to the world's human and nonhuman wonders.[28]

Institutions such as zoos were a vital part of these scientific and recreational provisions. Some late eighteenth- and early nineteenth-century traveling menageries (mobile animal collections that toured the country) tried to position themselves within these kinds of liberal and respectable parameters. Zoos tended, with some notable exceptions such as the Surrey Zoological Gardens, 1831–56, which was more of a fairground from the outset, to be much more inclined toward professional scientific endeavors and the provision of enlightened amusement.[29] When Sir Stamford Raffles established the Zoological Society of London in 1826, and its zoo in Regent's Park a short time afterwards, in 1828, he did so with a number of core aims in mind. The newly formed Society endeavored to educate, entertain, and display imperium at the heart of the Empire. It also imagined its animals as experimental subjects about which much could be learned and which could be molded and manipulated—acclimatized—for further economic use as domesticated livestock.

BRISTOL ZOO

Bristol's urban zoo emerged in this historical context, but it is a historical environment in possession of its own idiosyncrasies that render it distinct among the huge swathe of zoos that sprang up after the end of the eighteenth century.[30] It is the fifth oldest zoo in the world. It is also the oldest surviving provincial zoo anywhere, and while it certainly brought something of the exotic world to nineteenth-, twentieth-, and twenty-first-century Bristolian urbanites, its provincial character meant that it was rather removed from the intensely imperial and metropolitan concerns of London Zoo. As a local zoo, it was deeply entrenched in the city's own civic development. And yet it is also more than a local zoo, for it led the way in terms of zoo innovation and, later, conservation.

Inspired by the success of London Zoo, the Bristol, Clifton and West of England Zoological Society (today known as the Bristol Zoological Society) was created in July 1835 in a meeting held at the Bristol Institution for the Advancement of Science, Literature and the Arts, established in 1823 and which served as a nucleus of the city's empire of science. Chaired by Dr. Henry Riley, the Society's first honorary secretary, it was a private affair like many British zoological societies.[31] As private concerns, most zoos in Britain were not like those in the United States, which generally appeared after about 1870 and which were usually publicly owned. Around five hundred shares priced at £25 each were made available, and the monies generated were set aside to buy land, plant the Zoo Gardens, and procure the first animals. The initial members of the Society were among the local civic elite, but most were middle class, relying on an occupation of some kind for an income, rather than upper class. Predominantly, these men (and some women) were professionals, working particularly in the legal or medical professions. Many were also merchants and property owners.[32] It is no coincidence that many of the Society's founders were also involved in the city's Suspension Bridge Company (1830) and Philosophical and Literary Society (annexed to the Bristol

Institution, ca. 1820). The Society included men who were visibly successful across the sciences and industry, such as Dr. Henry Riley, a surgeon and lecturer in medicine who also dabbled in grave robbing in the late 1820s; Samuel Stutchbury, naturalist, geologist, curator of the Bristol Institution from 1831, and discoverer of the fourth-ever dinosaur found in the world (*Thecodonto-saurus*), in a quarry in close proximity to the Zoo; and H. O. Wills II, the son of a founding member of a renowned Bristol tobacconists company.[33]

Once financed, the Society's zoo was planted according to the extensive plans of Richard Forrest. He was former head gardener at Syon House, Middlesex, and had planted zoological gardens in Cheltenham and Manchester.[34] The Zoo was planted on an area of farmland known as Summer Trenmore on the outskirts of the fashionable north Bristol suburb of Clifton, an affluent area built on the profits of tobacco and slavery.[35] Many urban British zoos, such as London Zoo in Regent's Park, provided green spaces for fresh air and exercise, both of which were considered beneficial for body and soul. By the 1830s, Clifton was emerging as a popular destination for the middle classes looking to escape the increasingly crowded and polluted city center.[36] Situated on the northern edge of the suburb alongside farm and grazing land, the location was ideal for such a distraction from inner-city life.

Zoos like Bristol's formed part of a wide-ranging fabric of public and semi-public places emergent around the same period, which included gardens, museums, and art galleries. Reflecting the widespread social and scientific concerns of the age, the initial resolution of the newly formed Zoological Society was explicitly high-minded: "to promote the diffusion of useful knowledge by facilitating the study of the habits, forms, and structure of the Animal Kingdom, as well as afford rational amusement and recreation to the Visitors and Inhabitants of the neighbourhood."[37] Indeed, the Society was founded at a time when Bristol's own empire of knowledge was springing to life in the form of lectures and collections of natural objects like ferns, shells, and fossils of ancient beings.[38] The eighteenth century had seen the growth of the scientific community to such an extent that specialization in fields like geology, botany, and, of course, zoology characterized the early decades of the 1800s. The collection and study of these articles demanded the existence and perpetual invigoration of networks spanning the globe. The Society's shareholders, many of whom were themselves embedded within such arrangements, actively assisted in the procurement of animals and plants for the debut collection, for the gardens were designed to display botanical specimens, too.

The twelve-acre site included a central lake, a promenade, and a bear pit constructed on the site of a former limekiln. Once it was planted and animals were arranged in houses around its perimeter (fig. 1), the Clifton Zoo, as it was then known, opened in the summer of 1836.[39]

Although positioned in the affluent suburb of Clifton, the Zoo was never wholly isolated as a space servicing the social elites. Unlike London Zoo, the public was admitted from the very beginning, and this allowed it to become a place fully integrated into the social fabric of the city. Indeed, from the outset the principled ideals of the Society's founders had to be balanced against the demands of commercial viability. Visitors, and the money that came with them in their pockets, could drive forward the Society's higher scientific and benevolent agendas. This is not to say that the Zoo was wholly inclusive, of course: not least, the public was initially excluded from the gardens on Sundays, when the place was reserved for the exclusive use of shareholders and subscribers.

At around the same time, a wealth of zoological societies with similar aims emerged throughout the country. Unlike the majority of these, the Bristol Zoological Society survived despite a number of financially testing periods over the course of its first eighty years or so, and this makes it rather exceptional.[40] After the middle of the nineteenth century the Zoo transformed into a fairground of sorts in which it increasingly integrated into its landscape "amusements" such as fetes, fireworks, tightrope-walkers, dwarves, and giants. This was not so much a change of character as part of the solution to the ongoing financial troubles that plagued and then killed off so many other British zoos. In order to continue to educate and thus physically and morally build the nation (a rather elitist top-down approach to remedying perceived social ills), the Society had to be open to flexibility in achieving its ultimate ends.

In the twentieth century the Society prospered, particularly following the installation of Dr. Richard Clarke as secretary and Reginald Greed as superintendent in the late 1920s. It significantly reformed the character of its zoo, reverting to the more scientifically focused ideology of its earliest days, though fetes and fairs continued until 1946.[41] Meanwhile, Clarke's innovations, such as the 1928 faux-imperial Monkey Temple and the world's first Nocturnal House, in which daytime was transformed into night so as to allow visitors a glimpse of creatures who lurk in the darkness (1954), were complemented by a swathe of significant breeding firsts, such as the first black rhinoceros calf

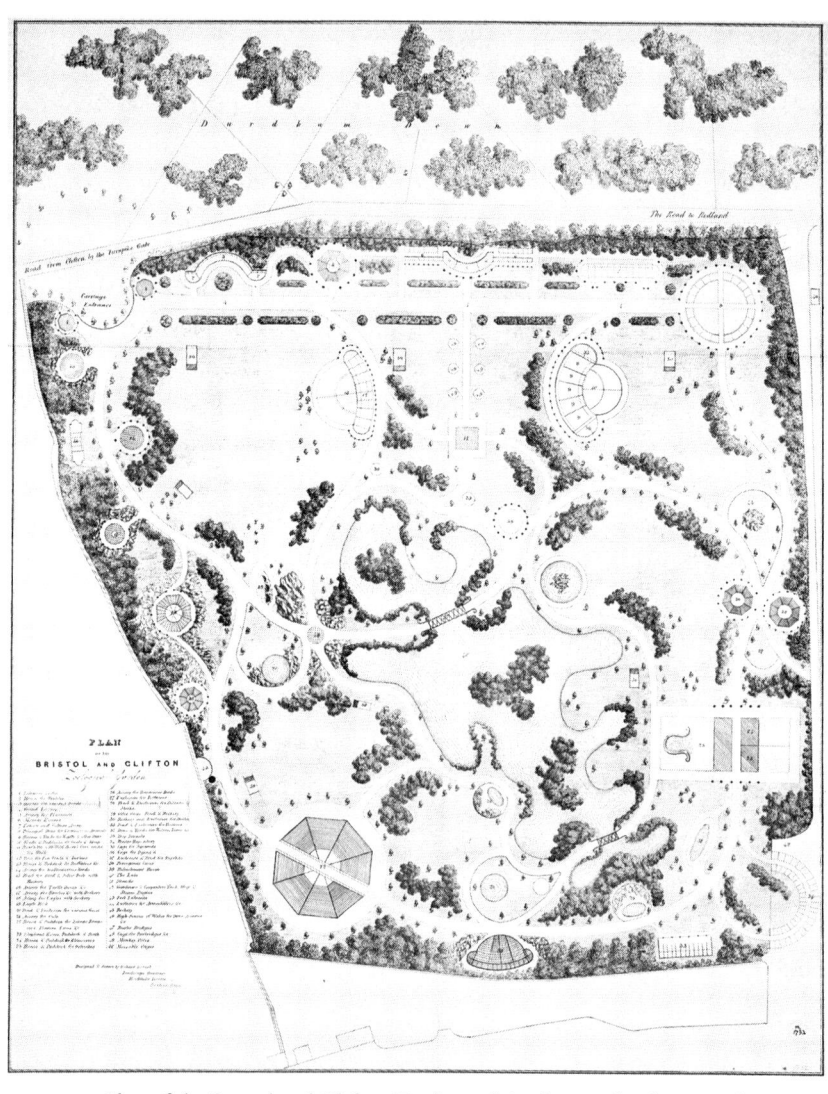

FIGURE I. Plan of the Bristol and Clifton Zoological Garden, Richard Forrest (c. 1774–1870). Lithograph on paper c. 1835. (Courtesy of Bristol Culture [Bristol Museums and Art Gallery M1732])

to be born in the UK, in 1958. At the same time, the cultural context of the day increasingly positioned animals as friends, as entertainers awaiting the arrival of visitors to the Zoo, and so, by the mid-twentieth century the Zoo had become a popular recreational space for families, generally affordable across classes. It had become less Clifton's Zoo and more *Bristol's* Zoo, reflecting its ever-broadening appeal and solidifying its position at the heart of Bristol's civic identity and leisure culture.

In the 1960s visitor figures were boosted by the arrival of zoological novelties, such as white tigers and the rather less popular but certainly odd okapi, as well as the generation of a succession of popular young. Importantly, too, the Society benefited from the democratization of the automobile, which increasingly made those with money to spare mobile in the pursuit of leisure activities, and the opening of the first Severn Bridge in 1966, which connected Bristol with South Wales and so opened up an entirely new population of potential visitors. All of this stimulated a significant increase in attendances, which passed the one million mark for the first time in 1967.[42]

These were heady heights indeed, when summer days were characterized by crowds struggling to move from one end of the Zoo to the other. From the late 1960s, however, the credibility of zoos was shaken as they were increasingly equated with animal exploitation. At the same time, the leisure industry exploded, offering more and more alternatives to a day out at the zoo. These factors, along with less disposable income resulting from the Depression of the 1970s meant that the Zoo suffered. Through little fault of his own, zoo director Geoffrey Greed (Reginald Greed's son) presided over falling attendances during the 1970s and 1980s. The Society was forced to significantly reinvent itself and its zoo once again. Very gradually, work in global conservation programs and conservation education moved toward the center of its identity. The extent of this reorientation is not clear-cut, but the arrival of its wholesome "green" image nevertheless stimulated a general recovery of visitor numbers, though it is still some way off reaching the heights achieved in the mid-1960s.

This "potted history" of Bristol Zoo demonstrates the general transformations that occurred within British zoos more broadly. This narrative, and the associated popular notion that zoos have forever been on some kind of upward trajectory, evolving, improving, culminating in their contemporary identity as island sanctuaries in a burgeoning sea of anthropogenic desolation, is, however, one-dimensional, even if it is a story that many zoological societies would love us to accept as an uncomplicated truth. In reality, zoos have always been places where all kinds of categorizations have collided and entwined, where a diversity of narratives have jostled for position. As hybrid spaces increasingly concerned with the past, present, and future in the forms of heritage, financial stability, and the conservation of a vulnerable world, zoos are fascinating places that are often seen as wholly artificial. That is, we imagine them to be of wholly human origin, worlds in which wild things simply and passively reside. Supposedly anathema to an idyllic imagination of a wild nature much removed from humanity's grasp, they are, as I have already suggested, far more complex than that, reflecting a diversity of relationships between humans and animals. At Bristol Zoo, across the entire span of its life, living beings were transformed into precious objects of science, commerce, and spectacle, as well as foci of human affection. Zoos have presented for the perusal of the inquisitive human gaze the loved, the loathed, and—increasingly—the (nearly) lost, alongside the docile and the dangerous denizens of the natural world, all the while constructing a complex imagination of a more-than-human biosphere. In essence, what they present—indeed, what they embody—are the many layers of human-animal relationship at the core of our experience of the world.

Aside from its local and global complexions, Bristol Zoo is a compelling case study. It is, as we have seen, the oldest surviving provincial zoo in the world and the fifth oldest anywhere. On top of that, its archive is unusually intact, and so an opportunity presents to study human-animal relationships across almost the entire lifespan of zoological gardens in Britain and beyond. It operates as a window into a hugely complicated and expansive world of human-animal relations. By examining the human-animal relationships therein, this book unveils a complex story of layering and entanglement. It sets out with two aims in mind: to illuminate the complexities, inequalities, and contradictions in the ways in which animals were construed, and to unveil the multifarious exchanges that took place between humans and animals in a hybrid but nonetheless often-anthropocentric world.

TRAVELS AMONG THE LAYERS, BY ELEPHANT

This book proceeds by foraging through layers of relationship, taking each in turn and charting its complexions over the course of modernity. It adopts this thematic approach not only to avoid undue repetition of key ideas but also in order to emphasize that there are many stories that one could relate about the highly complex world of the zoo and also about the often-unfathomable complexions of human relations with wildlife. Let us begin, then, by taking an elephant ride among these layers of human-animal relationship.

Over the course of its lifetime, the Society has displayed eleven elephants in its gardens, and many of these form integral characters in this work as we encounter them repeatedly in a diversity of material and imaginary inter-actions with people. The first appeared just a week after opening day in 1836 and was one of three acquired for exhibition and examination during the nineteenth century. Eight followed during the twentieth century, the last being Wendy, who died in 2002. A small selection of stories surrounding their lives among people in captivity demonstrates the way in which human-animal relationships were layered one on top of the next in a single hybrid space over more than 160 years. In so doing, I show that these layers were themselves complex, layered, and often characterized by hierarchy. Each layer, on their own and in combination, tells us much about ourselves, too, and the causes and consequences of our relations with wildlife over time.

Underscoring all of these relationships is an important question about the nature of the beasts themselves, and it is a question I posed at the very start of this book: *what are they?* When we think about human relationships with animals in captivity, we should be wary of considering these animals as pure—as wild—for, as we have seen, these terms are problematic indeed. Zoo animals, like the spaces they inhabit, are a particular kind of hybrid. Their very incorporation into the fabric of captivity, through capture/trans-location, display, and intensive management, transforms their natures, as Irus Braverman has convincingly argued. They are molded by their captive world, and so they are not the same as their free-living kin. Thus, these human-animal relationships are of particular complexions because the beings caught up in them are, to a substantial degree, the *products* of particular convergences of space, time, human and animal worlds.[43]

Whatever they may be, however, all of these elephants—every single one of them—were framed, often within a global context, as objects of desire. The first to arrive was an un-named male, procured by the Society's first "beast keeper" Harry Richardson for about £250 from a merchant who trans-ported him on the *SS Columbia* from Ceylon to the East London Docks.[44] Trumpeted as a "great addition to [the] collection," he arrived just over a week after the opening of the Gardens on 11 July 1836. For many zoos, elephants were intrinsic to the presentation of a *real* zoological collection.[45] The acquisition of animals was far from straightforward, of course. The capture and subsequent shipment of elephants to Europe was fraught with peril across the nineteenth and early twentieth centuries. During a long and increasingly anxious search for an elephant between 1910 and 1913, one was secured in Balasore, India. The Society had beseeched Eastern potentates for assistance, since it was unable to bear the cost of a full-grown elephant itself in the light of pressing financial concerns. He was moved to the Zoo over land on the Bengal-Nagpur railway, on foot, and then over sea on the British East India Company vessel *Nigaristan*. On arrival in London, and after a two-month-long voyage, he was discovered dead in his hold. The next elephant to be procured through similar channels and shipped at Calcutta drowned when the steamer transporting him hit a storm at Port Said, Egypt, in late 1912 or early 1913.[46]

For most of the Zoo's life, elephants were consistently objects of desire both for the Society and among its visitors. By the end of the twentieth century, however, their acquisition and maintenance had become problematic in the light of emerging tensions around the treatment of animal life. The Zoo's final elephant Wendy was extremely popular among zoo visitors in the final decades of the twentieth century. Despite this, an intensifying opposition to the captivity of larger animals in the late 1980s and early 1990s and the related emergence of conservationist agendas around the world and within and without captive environments amplified the already emerging conviction that she should be the Zoo's final elephant. When Wendy died in 2002, she was not replaced. These kinds of shifts in the extrinsic and intrinsic values assigned to and encircling animals form the basis of chapter 1. It shows that the Society consistently converted animals into objects of commercial and scientific longing. In so doing, it illustrates the Zoo's formative and exploitative position within the global wildlife market, while pointing to the multifarious ways in which animals were directly used in the generation of revenue. I trace fundamental changes and continuities in the value assigned to a diversity of species over time as British culture, and the Society itself, gradually and not always happily moved toward the reification of genetic material and a longing for natural heritage at the apex of "ecological modernity": an epoch Ian Jared Miller defines as witnessing the distancing of a mastered natural world and the animals within it from the daily rhythms of human lives.[47]

FIGURE 2. Judy giving rides, c. 1930. (Roy Vaughan Collection, Bristol Zoological Society Archive, courtesy of the Bristol Zoological Society)

All of these animals were acquired for a particular purpose, of course. On arrival at the Zoo they were "put on stage," transformed into spectacle. The Society's second pachyderm incumbent arrived on loan from the celebrated equestrian and circus proprietor William Batty for the summer of 1842. Known as Sambo Elephas, he was renowned for his extraordinary performances. In September, at the end of a busy season when the animal's part in the entertainment of curious visitors was almost up, the creature began to struggle during a dip in the lake. "Rescued from a watery grave" by Zoo attendants who dragged him to shore, he died of a ruptured heart two days later.[48] Nearly a century afterward two elephants, Rosie and the much younger Celeste, who both arrived in 1938, were among the Zoo's chief attractions. "The world's wonder elephant," star of circus, stage, and screen, Rosie was well known and well trained, and she formed a prominent part of the stories that follow over the course of this book.[49] Like some of the elephants who had preceded her, her body was deployed in the provision of rides for children along the Zoo's central artery: the Main Terrace. This kind of animal entertainment began in Bristol many years earlier, in 1869, and for a long time was an essential part of a child's visit to the Zoo (fig. 2).[50]

The practice of elephant rides was abandoned after Rosie's death in 1961, just as it was in many of the nation's other zoos. Health and safety concerns—

trepidation about the possibility of accidents—meant that wholesome spectacle in a family zoo no longer rested on these kinds of activities. Yet, the movement of elephants in and among visitors remained a central part of the zoo experience; Wendy and the elephant who lived with her for much of her time at the Zoo, Christina, were often walked around the Zoo Gardens and out along the Clifton thoroughfares. Thus, chapter 2 illustrates that captive creatures were not only conceived to be objects of a particular kind of scientific and commercial desire rooted in their possession. In many respects, they were valued precisely because they were seen as animate objects capable of captivating and delighting. Animals were not merely objects to be seen, however. They were also to be encountered in and around their captive worlds, and the nature of these encounters changed across the lifespan of the Zoo in ways that reflect changing notions of what nature—and wildlife—was supposed to be.

As well as objects of value in relation to the contexts of possession and of encounter, the bodies of many of the Society's elephants were also understood as resources in their role as vessels of knowledge that was to be extracted and effectively communicated. This layer of relationship was framed by an interest in the minutiae of animal bodies. The Society's first elephant lived in the Elephant Den for only a short time. By the end of 1838, it had died, and the valuable carcass was set aside for scientific examination. It was eventually donated to the Bristol Institution for dissection.[51] By breaking the body into pieces, much could be learned about the animal, not only in relation to its anatomical functions but also, potentially, in terms of how the species might be better maintained in captivity. The same fate awaited the body of the Zoo's final elephant, Wendy, when she died in 2002. The skeleton was dispatched to a research collection at National Museums Scotland, where it was preserved and its remains maintained for use in future research activities.[52]

This kind of knowledge was in itself extremely valuable. Large mammals like elephants required specialist care, and this mobilized the knowledge of the workings of animal bodies that was extracted via the kinds of activities exemplified above. For a long time this was provided by people local to the area from where the animal was captured. These people were not uncommon sights in nineteenth- and early twentieth-century zoos for the knowledge they brought with them was highly valued as a means of keeping treasured attractions alive. Indeed, they were often exotic attractions in and of them-

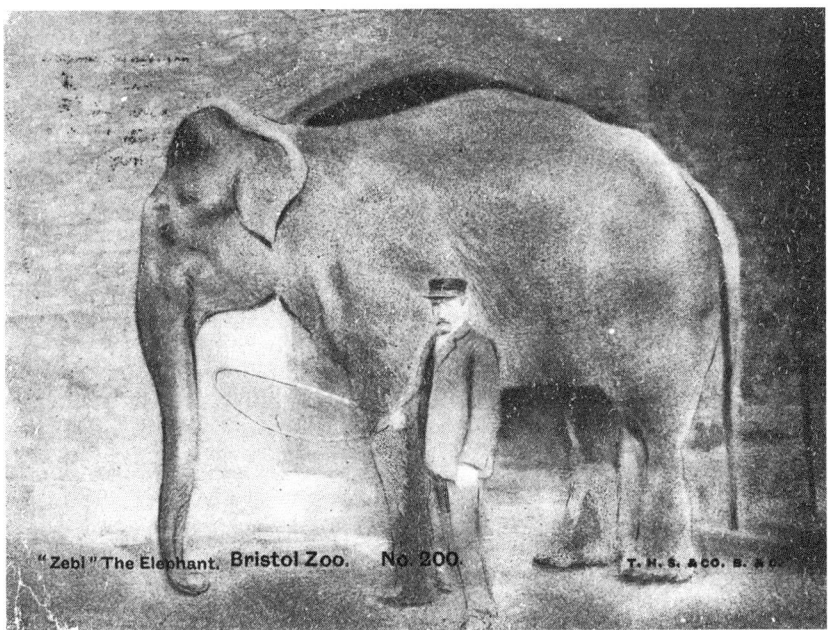

FIGURE 3. Zebi with keeper, c. 1870. (Roy Vaughan Collection, Bristol Zoological Society Archive, courtesy of the Bristol Zoological Society)

selves. Hamet Safi el Canaana, for instance, accompanied the popular hippopotamus, Obaysch, to London Zoo in 1849–50, while the "white elephant," Toung Taloung, was escorted to London by an "attendant" known as Radee in 1883. When an elephant named Zebi arrived in Bristol 1868, her mahout stayed until the Zoo's head keeper, George Blunsden, had acquired enough knowledge and established enough of a rapport with her to allow him to assume sole responsibility for her care (fig. 3).[53]

Chapter 3 thus examines human-animal relationships founded on the desire to know and exploit the deep secrets of nonhuman bodies. It delineates a widening and deepening comprehension of the dynamics at the heart of species and of the natural world and, subsequently, the ways in which that knowledge could be deployed to the end of practicing effective animal husbandry.

These three layers of human-animal relationship, rooted in the assigning of value through possession, the staging of encounter, and the desire to know and manage objectively, rested on a presumption of ontological distance, objectification, and humanity's inordinate ability to master nature. And yet, at the very same time as these processes of objectification were in full swing, all eleven elephants were also recognized and presented, to varying degrees, as subjects, with individual personalities of their very own. At least ten of the Society's eleven pachyderms were given personal nonscientific names, and they were popular as individuals who keepers and visitors could come to know as distinct characters. Chapter 4 focuses on a layer of relationship often configured by emotion and recognition. It illustrates that the manner by which animals were recognized as individuals within the mass of their species and the way in which they were, by extension, thought to be capable of experiencing the world, changed complexion over the course of the period. While Sambo Elephas was recognized as kind and docile, by the time Rosie and Celeste arrived in the late 1930s a cult of celebrity surrounding captive creatures was well established at a time when, culturally, many animals were perceived more widely as friends who actively craved engagement with people. Similarly, reactions to Wendy's demise in 2002 signify an enduring and highly developed conceptualization of this creature as an almost-human person. Her status as other-than-human was in flux in a way that elicited responses to her demise unlike any that had been seen in relation to Bristol Zoo elephants across the nineteenth and much of the early twentieth centuries. While her body was dissected in the name of scientific research, through acts of remembrance and mourning, her sentimentalized individuality endured among a pantheon of spectral beasts. Focusing at particular times on primates, reptiles, and fish, the chapter reveals a diversity of often intensely personal human relationships with nonhuman animals founded on fluid notions of sameness and difference. In so doing, it illustrates the diversity of incongruous ways in which humans tend to invest character in what they generically think of as the animal kingdom. Such relationships with captive animals were distinct from those that served to objectify and homogenize them. Yet both processes existed in tandem. Taken together, these layers denote the presence of human-animal relationships that were multifaceted and that appear to be, at their core, alarmingly contradictory.

Over the course of the first four chapters and as we explore four ambiguous and transforming layers of human-animal relationship, the Society and its visitors are largely revealed to be beings capable of a diverse set of manipulations of animal life. They controlled animal movements, mastered behaviors, and forged and imposed identities—both object and subject—in numerous ways. Despite all of this, however, human mastery of captive creatures was far from absolute. The Society's elephants were consistently influential actors in the forging of the zooscape and the human-animal interactions therein. Zebi proved, at times, to be difficult to manage, with a "mischievous" character. There were many stories "in vogue of her misconduct."[54] Judy, meanwhile, who arrived as a juvenile elephant in 1927, developed a taste for paint and, far more alarmingly for the Society, attacked senior keeper William Payne in 1931, fracturing his collarbone.[55] In addition, when Wendy and Christina were acquired in 1961, the two elephants were moved to temporary quarters in a potting shed. There they caused damage on an almost nightly basis.[56] Chapter 5, then, disrupts our notion of humans as all-powerful creatures and creators, particularly in captive environments such as zoos. It demonstrates that animals have been able to exert influence of their own, significantly impacting human behaviors, and in some ways shaping human-animal relationships across the period. In the zoo, both humans *and* animals had the capacity to create human-animal relationships, and the landscape in which they were performed, together. That said, animal actions were rarely left to stand for themselves. They were quickly interpreted or diverted. This was an interspecies relationship characterized by an ever-reconfiguring balance of authority.

In short, human engagements with the Bristol Zoo's eleven elephants over the course of nearly two centuries were profoundly complex. All of these kinds of relationships taken together show that the elephants were as much Wendy, Rosie, Celeste, or Zebi as they were *Elephas maximus* (Asian) or *Loxodonta africana* (African), a representative of a certain anthropogenic and anthropocentric conceptualization and evaluation of the natural world. Like the wild things who sprang to life in Max's bedroom, then, these were multifaceted, real and imaginary beasts dwelling in a place that was hybrid in nature and in which humans and animals often constructed each other and each other's identities in a shared world. In the pages that follow, I strip back those layers and explain how those strata of relationships were constituted over the course of modernity.

These layers were, of course, interwoven. We return to that at the end, in chapter 6. The point here, however, is to illustrate the incredible diversity, (in)consistency, and fluidity among ideas about animal life and many human relationships with the more-than-human world by considering in turn different manifestations of human-animal engagement—from commodification to negotiation. By using Bristol Zoo as a lens through which to examine these configurations, we might move toward an understanding not only of how our attitudes have changed over time but also how remarkably stable they have been. The past is clearly discernible in the present. Further, we see how layers of contradiction characterize our own attitudes and how this reveals something important about ourselves. In sum, stripping back these layers, one by one, may help us understand more clearly how we have come to relate to the natural world and its vibrant inhabitants in our age of ecological desolation. It might also help us to appreciate the wider consequences of those intricate relationships, not only in the here and now, but also in a delicate, uncertain, and interwoven future for human and animal life on Earth.[57]

1

Harvest and Heritage

Elephants were not, of course, the only star animal attractions. Big cats, in particular, quickly became well-established and familiar presences at Bristol Zoo. As soon as the Bristol, Clifton and West of England Zoological Society had been formed in the summer of 1835, the process of acquiring animals began in earnest, and one of the first specimens to arrive was a hunting tiger (leopard). It was kept by John Miller, who was a member of the Society and who housed the animal in his Clifton nurseries for the better part of a year until the Zoo was in a fit state to receive it.[1] Over the entire span of the Zoo's history big cats like this formed part of an elite set of *real* zoo animals; creatures whose presence made the Zoo worthy of the name. Indeed, up until the end of the twentieth century five or six species of big cat were on display at any one time. Typically these were multiple specimens of varieties of (usually Bengal) tiger, lion, leopard, and jaguar. The lion and tiger—whatever the variety—were particularly evocative symbols of the uninhibited power and ferocity of the natural world. Highly attractive to imperial hunters in the wilds of Africa and Southeast Asia, and frequently essential to the iconography of imperialism and nobility, they have been at the heart of exotic animal collections since their emergence.[2] At Bristol, the value attached to their possession was elevated well above most of the rest of the zoological collection. In some circumstances, expenditure was secondary to the necessity of acquisition: in 1838, the Society secured a bank loan of £150 specifically for the purchase of a lion; these funds were intended to contribute to the overall cost involved in procuring and then transporting the creature, which, it was very clearly stipulated, was *not* to exceed £210. In any case, the final cost was £250; the Zoological Society of London would not accept a penny less, and the Society gave in. A similar transaction took place in 1853 when the Society sought a lion from the renowned animal dealer Charles Jamrach, whose notorious animal emporium was located on the Ratcliffe Highway in London's East End. The priority was the lion and not the expense incurred, the Zoological Society having resolved "at all events to secure it."[3]

By the end of the nineteenth century, the Society had earned a national reputation for possessing superior (i.e., larger, physically attractive, glossier manes, in the case of lions at least) big cats in comparison to zoos elsewhere in the country.[4] Local newspapers habitually reported the arrival of new animals, for these kinds of acquisitions were suffused with popular appeal. In 1890, for instance, the *Bristol Mercury and Daily Post* reported that the procurement of two tigers rendered the collection of carnivora "one of the finest in the Kingdom," if not in Europe. Later, Florence Grinfield, who was the author of an 1894 unofficial guide to the Zoo, declared that the admission price of sixpence would remain worthwhile if nothing but lions were on display.[5] Their perceived status among the various denizens of the animal world was clear for all to see.

They were extremely valuable creatures, but they were not merely objects to be acquired for display. Their bodies could preserve, and indeed mature, a latent financial value, which could be cashed in at a moment of the Society's choosing. In 1894, for instance, the Society decided to sell one of its Bengal tigresses to the American circus showmen Barnum and Bailey, in a transaction that was worth £125 and that, again, included the services of the great London animal dealer Charles Jamrach. Restrained within a wooden crate and transported by passenger train between Bristol and London Paddington station on Isambard Kingdom Brunel's Great Western Railway, the animal was then transferred onto a ship for the lengthy journey across the Atlantic to the United States, where it was destined to deliver a magnificent spectacle to the patrons of the showmen's "Greatest Show on Earth."[6] Through this transaction, the Society cashed in a valuable commodity by reinserting it into the same global trading arena through which it assembled the greater proportion of its animal stock. Animals did not simply flow ever inward to Bristol

FIGURE 4. White tigers, c. 1963. (Roy Vaughan Collection, Bristol Zoological Society Archive, courtesy of the Bristol Zoological Society)

Zoo. Instead, they circulated the globe in a complex interconnected multidirectional market system that decisively framed animals within a global context as objects of desire, as resources to be harvested from across the world's bounteous habitats.[7]

As objects of desire, not all animals were equal, however. Big cats remained rested at the peak of a hierarchy of animal value, and there were few capable of dislodging them. In 1963, however, when the Zoo was nearing the pinnacle of its popularity, Bristolians welcomed an extraordinary pair of cats to their Zoo. They were quite unlike anything anybody in the city had ever seen before. The curious beasts lured visitors to the Zoo in droves. In the process, they earned themselves and their captive institution quite a reputation on the national and international stage, for there were no creatures like them anywhere else in Europe. They were two white tigers, descended from an individual known as Mohan, who had been captured in 1951 in the jungles of Govindgarh near Rewa, Madhya Pradesh, and were bought from the Maharaja of Rewa for the princely sum of £6,400. That is the equivalent of about £97,000 in today's money and compares to £1,200 the same year for a normal-colored Bengal tigress (fig. 4).[8]

The exhibition of the tigers boosted visitor figures from 796,437 in 1962 to 874,263 in 1963 and a record-breaking 894,055 in 1964. Five years later, and after a couple of failed attempts, four tiger cubs were born. According to one Australian publication (for their fame was global) they were "worth five times their weight in gold." A film produced by British Pathé remarked that the cubs were "behaving as though they were born destined for great things—and they are—the four white tiger cubs . . . [are] real rarities . . . priceless in fact."[9]

By the time of the cubs' birth, the Society held one third of the entire global captive population of the creatures. Indeed, they were so commercially lucrative that by 1970 each tiger was valued at an astonishing £16,400 (about £170,000 in today's money).[10] Coordinated breeding efforts in the early 1970s sought to preserve the "white strain" to secure a self-replicating supply of the prized creatures.[11] The possession of these precious beasts remained tightly bound to the Zoo's identity well into the 1980s. The final decades of the twentieth century, of course, bore witness to a transformation at the very core of the Society and its zoo. The confluence of changing cultural, economic, and scientific currents remolded the way in which many, but not all, animals were construed as objects of desire, and this eventually led to adaptations in the Society's stock policy. An increasingly ethical and largely conservation-based ideology came to underpin the rationale of acquisition, dramatically affecting the position of big cats at the Zoo. The white tigers were cast out of the collection in 1985. Research into the genetics of the animal revealed that it could not be considered to be a subspecies at all but was, in fact, the product of a genetic mutation that was associated with deformity and high mortality rates among young animals. Consequently, they had no conservation value at all. Once their genetics were exposed as the cause of much suffering, these prized and commercially lucrative animals were no longer considered objects of desire, and they lost their place in the collection.[12]

Over the course of the next two decades, the five or six species of big cat that had been at the heart of the collection for much of the previous century and a half were whittled down. At the beginning of the twenty-first century there was a single species at the Zoo—the Asiatic lion—which was made to stand for all big cats. With merely some five hundred individuals remaining outside of captivity in the world, the presence of these critically endangered animals reflects both the conservation agenda, which is now so influential across many of the world's zoos, and the consideration of some species as planetary heritage in need of protection from destructive anthropogenic forces.[13] Importantly, however, the Asiatic lion is still *lion,* and it is still king of the zoo. Positioned just beyond the main gates so that it is one of the first animals encountered upon entrance, it is hard to avoid the sense that the animal continues to hold its place among an elite set of ultra-precious zoo animals. Deployed by the Society as representative of all big cats, the animal serves as a flagship species in the conservation age. Compelling and capable of inspiring wonder, the popularity of the animal draws attention to the plight—and intrinsic value—of the natural world at large.

The history of big cats at Bristol Zoo is more than merely a window into the changes and continuities at the heart of the Zoo, as it has existed for the best part of two centuries. It is also about the shifting contours of a fundamental relationship between animals and people and one that often takes place on the largest of scales: our mastery of them, and their conversion into homogenized commodities whose exchange connects distant worlds. This chapter is about that important layer of relationship: perceptions of the value ascribed to nature. This relationship, rooted in objectification and the crafting of speciesist hierarchies, was at the heart of Bristol's historic zoo. In fact, it underpins so many of our relationships with the natural world. Many of these cats—these "lively commodities"—were among an unimaginable multitude of beings lifted from their habitats and moved across land and sea in order to complete the zoological society's mission of accumulating a scientifically and commercially valuable zoological collection.[14] In the process, they moved in multiple directions in a global trading arena comprising myriad market terminals, and for a long time finance followed in their wake.

Over the past two hundred years—an era of accelerating extractive and consumer capitalism—the intensifying objectification and commodification of animals has been a fundamental interspecies relationship. This relationship was embedded in the imagination of animals as objects of some tangible worth. The wild animal trade, which was a manifestation of this attitude and which facilitated the accumulation of animals in Britain's zoos, operated in a reasonably consistent way over much of the course of the nineteenth and twentieth centuries. However, the vast international webs through which animals and finance flowed had a vividly different complexion by the end of the twentieth century. These changes directed, and were directed by, transforming attitudes toward nature, not least its rising characterization as vulnerable in the face of human industrial and commercial manipulations. These changes, however, were not without their tensions. Increasingly, the worth of animal bodies was perceived to be not *only* rooted in the extrinsic scientific and financial terms that were, as discussed, attached to the bodies of big cats, but also in their intrinsic value, the supposedly essential nature of their being. In an era characterized by an increasing awareness of ozone depletion, catastrophic habitat depletion, and accelerating climate change, this intrinsic value came to reflect the emerging sense of relationship between healthy genes, healthy species, and vibrant environments, and it is precisely this holistic relationship framed on a planetary scale that has become of immense value.[15]

What is without doubt a substantial shift in the broad perception of animal value across the nineteenth and late twentieth centuries does not mean that there was a straightforward transformation over the course of the period. The disparities between valuations at either end of the period only tell a fraction of a complicated multispecies story. In fact, human relationships with wildlife rooted in objectification and in perceptions of value have changed but have also remained astonishingly consistent. The binary imperatives of science and commerce that underpinned perceptions of animal value endured in unison into the age of zoo-based conservation. Similarly, the human sense of superiority over—and separation from—the rest of the natural world remained robust across the period, even if somewhat altered in complexion. Indeed, lurking amid all of the value-based imaginings of wildlife that follow in the pages of this chapter was an enduring faith in human exceptionalism, whether it was through the justifications underpinning exploitative extraction and exchange, or in the context of efforts to secure species survival.

MINES OF ZOOLOGICAL WEALTH

Over the course of the history of human relationships with other animals, their extensive and diverse objectification as tangible resources has been central. Through the transformation of animals into biocapital, a definitive and enduring distinction has been drawn, separating animals as property from their human owners and testifying to human power over the rest of the natural world. Wild things in our midst have been chased down and then slaughtered, their skins and pelts manipulated to form items of clothing and decoration, while their fleshy and bony bits—like tiger, deer, turtle, and oxen penis—have been harvested for use as food and medicine. These uses reflect some of the countless ways in which people have turned living things into objects whose extrinsic value resides in their physical relation to human concerns. As part of this, living creatures have been transformed within institutions of learning and amusement. In zoos all over the world, they have been transmuted into simultaneously commercial and scientific specimens whose presence alone was vital to the generation of revenue and the accumulation and dissemination of worldly knowledge.

From its inception, and right through the nineteenth and much of the twentieth centuries, the Society's relationship with the natural world at large, and specifically with the animals who inhabited it, was shaped by its desire to acquire and maintain a collection that fulfilled its dual functions of science and spectacle. This meant that human-animal relations were configured within a frame in which living things were converted into tangible commodities. These purposes, however, were not of separate significance. A scientifically oriented collection and presentation of specimens could only be achieved through the possession of a collection that was also capable of enticing paying crowds of the curious. Within this system, the values attached to species varied considerably. In one measure, individual animals were conceived of as specimens, as representative and equal parts of a larger catalog of nature. At the same time, rare or popular creatures, and young versions of both, were elevated above the rest as especially desirable commodities capable of both enriching the scientific collection and enticing visitors through the gates. Importantly, too, those that fell short of the ideal of perfection so crucial to the fulfillment of both the Society's scientific and commercial ends were cast aside to be replaced by specimens more acquiescent to the required aesthetic standards.

The Society's involvement in the global wild animal trade across the nineteenth and early twentieth centuries testifies to these conceptualizations of animal life. When the Society was established in the late summer of 1835, it immediately began to tap into imperial mechanisms of extraction and translocation. By 1843 (the date of the earliest surviving *Catalogue of the Animals*) the Society possessed creatures lifted from environments the world over. The core of the collection was of sub-Saharan African origin, though early exhibits also included gold and silver pheasants from China, varieties of primates and cats from India and Southeast Asia, a range of marsupials from New South Wales and New Holland, birds and boas from South America, and moose and Labrador dogs from North America.[16] A reliance on sub-Saharan Africa and the West Indies is unsurprising given the relatively short journey times, ease of access, and Bristol's continuing dependence on well-established commercial and familial links with West Africa, America, and the West Indies as a legacy of slavery and its various commercial outputs such as sugar, tobacco, and palm oil.[17] Such an expansive collection was made possible through the Society's involvement in a vibrant animal trade. During the rapid expansion of European empires and the concurrent rise of industrial worlds over the course of the eighteenth and nineteenth centuries, the ability of industrialized societies to extract natural resources and to move them from place to place increased. The scientific and commercial uses that could be made of the world's materials came to dominate ways of viewing and evaluating territories beyond the boundaries of Europe that were being harvested for resources, including living creatures, at an accelerating pace. Such inclinations, growing technological proficiency, and the development of increasingly efficient lines of global communication and translocation (transport around the British Empire was rendered immeasurably more efficient by, for instance, the opening of railways in India in the mid-nineteenth century and the culmination of the construction of the Suez Canal in 1869) not co-incidentally corresponded to the emergence of a significant escalation in the number of public and private animal collections across Europe. Indeed, these nineteenth-century transformations stimulated—and were stimulated by—the "natural history craze" during which the popular collection of shells and mollusks, fossils and ferns as well as living (and dead) creatures, intensified.[18]

The consideration of animals in this way meant that the possession of lively animal objects was what mattered most in the nineteenth-century zoo. The process of acquiring these precious commodities was astonishingly wasteful. The memoirs of Carl Hagenbeck, an eminent German international dealer and showman whose animal and ethnographic business operated from the second half of the nineteenth century, reveals not only that entire herds were run to their deaths for the sake of ensnaring just a handful of specimens, but also that the practice of slaughtering powerful adult animals in order to seize their vulnerable young was commonplace. Even once captured, animals regularly died of exhaustion later, having endured an often long and arduous chase.[19] In early 1893, for instance, Bristol Zoo lions Libah and Juba were successfully captured but only after the slaughter of their parents in the forests of what was then known as the protectorate of British Somaliland.[20]

The death toll continued to rise, even once wildlife had been apparently brought under human control through capture. The ferocity, fragility, and edibility of animals (for many of them proved to be irresistibly mouth-watering to hungry sailors) meant that creatures loaded aboard ships for the passage to England did not always reach their destinations alive. Confined in small cages and tossed around at the will of the waves, big cats, for instance, "tore themselves to ribbons and bled to death, or put their own eyes out," while poor-quality food and water impacted on the ability of animal bodies to endure grueling journeys.[21] The scale of maritime mortality of animals bound for Bristol Zoo in the nineteenth and early twentieth centuries was substantial. In 1863, a Captain Hassell departed from the west coast of Africa with a consignment of at least seven monkeys, a chimpanzee, two porcupines, and a civet cat, all of which died on the voyage to the city, while in 1890 two prized tigers died en route from Jeypore to Bristol.[22] Yet, they kept coming on a perpetually rolling transcontinental conveyor belt laden with precious animal things.

If levels of mortality implicated in processes of capture and relocation were largely hidden beyond the Society's gaze, they were made abundantly clear once animals had arrived at the Zoo. In captivity animals died in immense numbers. The cold, damp conditions of British winters and the transmission of disease were significant problems in the maintenance of animal life. Indeed, in 1841, maintaining the vitality of animals through an entire winter was dubbed "a great experiment," clearly illustrating the perils inherent in the maintenance of life. Indeed, this great experiment in animal husbandry proved to be largely unsuccessful over the course of much of the Zoo's first century (see chapter 3). The period 1854–55 saw the "less numerous than usual" number of twenty-three mortalities, while 1855–56 witnessed the demise of sixty-six animals. The maintenance of primates was a particular challenge, given their vulnerability to respiratory disease in the damp and chilly British climate. The 1843 *Catalogue of the Animals* reported that very few survived for more than a few months. Indeed, in 1881 three-quarters of the Zoo's monkey collection died during a particularly severe winter.[23]

Mortality itself was the driving force behind the global wild animal trade.

Replacements were unendingly sought, while continuing exploration perpetuated a view of the world as immeasurably large and teeming with life.
During the nineteenth century, despite alarming animal depletions increasingly in evidence at the Southern African Cape as the century wore on, environments (and thus animal populations) were opened up as Europeans ven-
tured deep into the African interior.[24] The vast sub-Saharan expanse of what
was in the nineteenth century often labeled as "the Dark Continent" and the
inaccessibility of forested areas such as those of Western Africa, as well as
the nineteenth-century intensification of the mass extraction, or harvesting,
of natural resources like palm oil from the continent, might have enhanced
a view of African animals, in particular, as plentiful.[25] Certainly, in 1851, the
Society referred to the west coast of Africa as an "almost unexplored mine of
zoological wealth," the use of the term "mine" inferring a notion of wildlife
as plentiful, if not superabundant.[26]

The degree of destruction wrapped up in the nineteenth- and early
twentieth-century acquisition and maintenance of animals is important. It
testifies to a conceptualization of much of the natural world as something of
an "open air treasure-trove" across this extended period and to human-animal
relationships rooted in that sense of abundance and a philosophy of perpetual objectification. The impact of continuous extraction on environments
and species was, for a long time, thought to be negligible given the incon-
ceivable quantity and vibrancy of wildlife that was imagined to be available
for almost endless acquisition.[27] This, combined with the sheer numbers of
animals arriving at European ports, demonstrates that the Society essentially
considered animals to be "easily renewable" resources.[28]

This sense of superabundance aside, high levels of mortality meant that
the Society was compelled to cultivate reliable, and diverse, supply lines
to ensure a consistently comprehensive stock of animals. During the nineteenth century, animals entered Bristol through various market streams after
circulating in an expansive, though largely informal, global trading arena.[29]
As with other European—and indeed global—zoos, there was no preferred
provider; instead, numerous individuals and organizations were connected
rather haphazardly in the supply of animals for purchase.[30] Lively commodities were procured from professional animal dealers such as the London-
based Charles Jamrach (and, later, his sons Albert Edward and William),
J. D. Hamlyn, Charles Rice, William Cross of Liverpool, and Hamburg's

Carl Hagenbeck.[31] So vital were the Society's connections to such suppliers that, in 1891, Secretary Frederick W. Field Collier traveled to London specifically to foster the Society's relationships with the capital's prominent animal dealers.[32] In addition, animal stock was acquired from other zoos, circuses, traveling menageries, and private individuals. A large male lion called Hannibal, for instance, was acquired in an 1872 auction when Wombwell's Royal No. 1 Menagerie was disbanded.[33] Members of the Society itself—many of them being merchants with extensive international connections—were also active in acquiring animals during the nineteenth century. Early on, it was noted that "the Committee congratulate the Society on the great willingness of persons having foreign connections to procure animals for the Society." Wine merchant William Wheeler Green was one such individual, regularly procuring reptiles and amphibians in the 1850s and 1860s via his contacts in the West Indies.[34]

Beyond monetary exchange through the strictly commercial channels of the wild animal trade, animals were also gifted to the Society through the well-established conduits of a global network characterized by patronage.[35] In this regard, the conversion of animals into tangible commodities was complicated by their simultaneous consideration as symbols of support and friendship. Bound up within the animal commodity was a political or sentimental statement. Over the course of the nineteenth century, diplomats, foreign dignitaries, and other zoos gave an average of sixty animals a year to the Society as presents. Mariners were especially reliable sources of such favors. Between 1850 and 1893—the peak of this particular mode of acquisition—seamen landing at Bristol made at least 108 donations, each often comprising multiple animals.[36] In addition, household animals, particularly monkeys and birds, were frequently converted from objects of emotional attachment into zoological specimens through their donation to the Zoo. Such donations were actively encouraged to avoid the Society's unnecessary expenditure across a century during which its finances were not infrequently precarious. In the 1830s and 1840s, at least, donors were rewarded with complementary admission tickets, sometimes covering a period of up to a decade. Others received commemorative items of silverware. Significantly, too, such gifts were credited publicly right across the century, thereby heightening the prestige associated with donating live animals to the Society and encouraging the perpetuation of this practice.[37]

A HIERARCHY OF PRECIOUS THINGS

While these attitudes and the transnational systems in which they were reflected transformed animals into assets that could be owned and moved from place to place, the values assigned to species were not at all uniform. Objectification through extractive activities was itself a complex human-animal relationship. The significance attached to species was nuanced by a number of competing and complementary scientific and commercial imperatives. Consistently throughout the nineteenth and early to mid-twentieth centuries, *all* animals in zoos were at one level equally construed as representative scientific specimens. As if objects arranged in nineteenth-century museums, these zoological specimens were stationed in a comprehensive and anthropocentrically determined series that facilitated the inspection of an orderly and thus comprehensible—to natural scientist and visitor alike—animal world. Such an ethos could be found across nineteenth-century learned institutions. The opening address of the January 1835 *West of England Journal of Science and Literature* proclaimed that the Bristol Institution held a collection that was "a large and respectable one, containing nearly complete Zoological, Conchological and Mineralogical series." Similarly, a principal purpose of Richard Owen's 1862 plans for a National Museum of Natural History at South Kensington, London, was "in order to lodge the samples of the Works of Creation of every class, of all time, and from the whole World."[38] In these places of scientific endeavor, the possession of all was the aspiration, and specimens, be they animal or otherwise, were equally valued for the constituent roles they played in populating such a holistic scientific collection capable of enhancing urban society. At Bristol Zoo, the display of many and varied species was intrinsic to the validation of the Society's claim to possession and exhibition of a zoological series, rather than merely a sporadic array of curious and popular species akin to the traveling menageries of the age.[39] The Society's 1854 *Annual Report* notes, for instance, that "your Committee . . . lost no time in filling up the blanks in your Menagerie as they occurred, adding more specimens where possible."[40] Native as well as exotic species were essential to the composition of an all-inclusive collection. An aviary for English birds formed part of the exhibition from 1863, and by 1899, Aviary No. 2 contained specimens of chaffinch, siskin, brambling and hedge sparrow, all species that could easily be spotted in the trees and hedgerows of the British Isles. A number of English foxes and squirrels were also displayed at the Zoo across the nineteenth and early twentieth centuries.[41]

On the face of it, the objective of forging such a supposedly representative collection implies that each animal had its designated, and *equal,* place in some grand conceptualization of animal life. Yet it is clear that, as George Orwell's porcine Napoleon memorably put it, some animals were "more equal than others."[42] Competing and complicating commercial and scientific imperatives meant that some species were positioned above others in an anthropocentric and speciesist hierarchy of precious things. Compiling and then maintaining a comprehensive zoological collection did not come cheap. The attraction of fee-paying visitors thus played a vital role in the formulation of stock policy. For this reason, animals that were well regarded by visitors and that were thus popular attractions—for whatever reason—were also especially esteemed by the Society. In 1839, for instance, the Society attempted to acquire an otter, a species that had proved to be exceptionally popular at London Zoo. Often hidden in the undergrowth or beneath the waters of native wilds, the species was imbued with a sense of the exotic. The Society's decision to try to capitalize on its popularity as an exhibit illustrates their desire to display creatures that were not only part of a comprehensive collection but that could also heighten the reputation of the Zoo as an attractive place to visit.[43] Such commercial and scientific valuations of animal life were certainly not mutually exclusive. In many instances, these layers of esteem interlocked. Rarely seen creatures were highly esteemed both as critical components of a comprehensive zoological series and for their ability to attract and intrigue visitors. The more unusual the species, the more commercially valuable it was as a means of luring visitors through the gates and the more vital it was to an aspirationally comprehensive collection.[44] In 1864, the Society explicitly stipulated that it wanted to improve accommodation for any "rare and valuable" species. Such creatures were evidently ascribed exceptional value, which in turn required provisions that would not usually be afforded to the rest of the collection.[45] The esteem attached to zoological curiosities perpetuated a steady stream of novelties flowing into the Zoo over the course of the century. The first two-toed sloth ever displayed in Britain, for instance, was received with much acclaim in 1850, and in 1860, the Society unsuccessfully sought to acquire a hippopotamus, probably having been influenced by the metropolitan celebrity of Obaysch at the London Zoo during the "Hippomania" that he inspired in the early 1850s.[46] Some years later, in 1892, when a White fox (Arctic fox) was donated, the *Bristol Mercury and Daily Post* implored readers to visit the Zoo to catch a glimpse of such a splendid rarity.[47]

Great apes sit neatly at the convergence of scientific and commercial valuation during the nineteenth century, for they were rarely seen in zoos well into the twentieth century. Some of the earliest European descriptions of part-human, part-animal monstrosities such as the "engeco" (chimpanzee) and orangutan date only from the seventeenth century, while the gorilla (formerly known as "pongo") was only fully described in the 1840s. These species were far from familiar to the European eye.[48] Furthermore, great apes were rarely maintained successfully in nineteenth-century zoos, and this enhanced the whiff of novelty that permeated their presence in zoological collections. In the early years of the Zoo, chimpanzees and ungka-puti (agile gibbon) not only constituted essential elements of the representative, comprehensive scientific collection but also were compelling as rarely seen specimens and commercially lucrative curiosities for public inspection. The first chimpanzee to be displayed at the Zoo arrived in July 1838 and could be viewed for an additional payment of one penny per person while, in June 1839, July 1840, and April 1841, ungka-putis were put on display.[49] The potent convergence of science and commerce in the aesthetic of rarity attached to great apes is particularly striking in events during 1841 when the Society paid £100 for a pair of chimpanzees. The creatures were so popular that they were also displayed at the Grand Dahlia Show in the nearby city of Bath, and negotiations were begun about their exhibition at the Cheltenham Zoological Gardens, though their anticipated physical fragility in the winter months put an end to that particular venture. Strikingly, such was the value attached to the possession of this species that the Society continued to publicize its former presence in guidebooks long after the animals had died.[50]

The value assigned to such novelties was usually impermanent, however. Creatures once highly valued as exotic curios became less valuable as the aesthetic of the exotic that was once so deeply inscribed upon their forms faded away. The transformation of animal value in this regard is striking in the context of the Zoo's nineteenth-century exhibition of dog breeds.[51] Around the mid-nineteenth century labradors, St. Bernards, Alaskan malamutes, Macrow wolf dogs, and Russian sled dogs were all part of the zoological collection. While these creatures were part of the Society's representative series, they were also unusual enough in Britain at the time to constitute exotic, indeed rare, subspecies. Labradors, originating on the east coast of the Labrador region of the Canadian province of Newfoundland and Labrador, were only introduced, in a domesticated sense, into the United Kingdom in the late nineteenth century. St. Bernards, meanwhile, had supposedly originated near the St. Bernard's Pass, Switzerland, where they gained legendary status for rescuing stranded travelers along the treacherous mountain pass. Alaskan malamutes, on the other hand, were generally associated with the inhospitable places of the world and the exotic lives of the local peoples of the Arctic.[52] By the end of the nineteenth century the vast majority of these animals had disappeared from the collection, their value in an exhibition of the exotic having been diminished through their conversion from rarely seen creatures into increasingly familiar companions. While this process illustrates that the wild and the tame exist along a continuum, where one can transform into the other, it also reveals that the values assigned to lively commodities were inherently fluid.[53] It was for precisely this reason that new species were regularly required in a process that Harriet Ritvo so neatly describes as a "boom and bust" cycle that sought to stimulate visitor interest over and over again through the regular presentation of the perplexing and the rare.[54]

With novelty value often so impermanent, species considered particularly charismatic—consistently popular and capable of drawing crowds—were especially prized by the Society over the course of the nineteenth century, for they could almost guarantee the generation of revenue through ticket sales. Quintessential large zoo animals, charismatic megafauna such as big cats, polar bears, and elephants, were especially valuable to the Society's commercial operation. Neither merely equally valued components of a representative zoological collection nor particularly rare curios, these species were viewed all over the world as essential and familiar constituents of any zoo.[55] Certainly, a large portion of the appeal of these charismatic megafauna may have been their prominence in popular culture.[56] Lions, tigers, elephants, and zebras, for instance, featured conspicuously in the traveling menageries of the day and in popular yarns of exotic worlds such as the nineteenth- and early twentieth-century great white hunter narratives and boys' adventure fiction such as W. H. G. Kingston's *In the Wilds of Africa* (1871). The popular appeal of these species ought not to be underestimated. In 1858, the children of a workhouse in the Bristol district of Stapleton visited the Gardens and were moved at being "in the presence of real live ... [animals], of whose existence they had only previously known through the medium of pictures."[57] Such contexts were the means through which expectations of the wild places and wild things of the world were set in the absence of widely available foreign travel.[58] Having these kinds of animals on display was crucial to the construction of an aesthetic that evoked a sense of wildness: alterity and human mastery.

The Society's nineteenth-century procurement of polar bears reflects the elevated value assigned to these varieties of charismatic species rather well. The death of the Society's polar bear in 1881 was the subject of a newspaper article that suggested that he was one of the Zoo's most popular animals.[59] Such was the reputation of this species that the arrival of another bear in 1894 from shipping agents in Dundee generated press interest, and when the Society again found itself without a polar bear in 1905, it issued an urgent call to "friends" who might wield influence with the North Sea fisheries or whaling companies asking for them to produce a polar bear at once. These shipping and fishing agents routinely encountered the bears on the ice floes of the North Sea.[60]

The high estimation of charismatic species is clearer still in the prices they commanded in the wild animal trade. Between the mid-1860s and the early 1880s, for instance, such species regularly attracted much higher monetary values than did others. Big cats, in normal circumstances, commanded upward of £50, while a polar bear usually cost about £30. In comparison, most animals added to the collection were valued at around £5, at most, for an individual specimen, and as was typical among European zoos, monkeys and many small birds often cost far less. Although these prices were often informed by levels of supply in the marketplace, they also reflected species popularity: the pricing of small monkeys and many birds bears witness to their ongoing ready availability in the market and their relatively lesser popularity compared to large and rare beasts.[61]

Beneath the level of species, nineteenth- and early twentieth-century variations in animal value were complicated still further by consideration of the age and sex of specimens. Both of these could heighten or lessen the desirability of lively commodities. Young versions of the Zoo's most popular species were especially popular with the Zoo's visiting public, and this rendered them of increased value to the Society. In 1854, it was reported that "the public will learn with interest that the attractions to these gardens are considerably increased, from the circumstance that the young lion cubs, with which Madam Juno lately presented her liege lord, are now on view every day. . . . The exhibition of these graceful creatures will doubtless largely increase the numbers of visitors."[62] So great was the Society's experience of public excitement upon the event of a new birth that in 1904 it explicitly stated that "we are hoping, of course, for favors to come, and if they should assume the shape of young tigers, a baby elephant or a polar bear cub, they will prove very especially acceptable."[63] Accordingly, the value of young animals as popular exhibits, combining both charisma and transience, simultaneously attached high worth to individuals who would be most likely to produce them. Fertile females were thus high on the Society's shopping list. In 1859 a fresh lioness, a jaguar, and leopards were sought to replace aging specimens; it was hoped that the younger versions would have a better chance of successfully producing cubs. Similarly, in 1869, a litter of lion cubs failed to mature, it was thought, because of the advanced age of their mother. The Society thus sought out a younger lioness in an effort to generate cubs. Triumph in the breeding stakes was not merely a matter to be pinned on the state of the female of the species, of course. In 1890, Secretary John Tom Jackson was actively seeking to acquire a male lion, specifically "for stud purposes."[64]

The production of a steady supply of young and popular animals signaled a determination within the Society to perpetually "reproduce" the Zoo by generating desirable species, and finance, for itself.[65] If it could ensure the continued fertility and productivity of adult stock, then it could equally ensure the continued supply of the animals that could most effectively address its dual commercial and scientific requirements. The propagation of a self-supporting supply mechanism also served the purpose of saving the Society significant expenditure; as we have seen, the animals most likely to draw in the crowds were also often the most expensive to acquire through the channels of the wild animal trade.

The worth of these variously valued commodities, whatever their species, sex, or level of fertility, relied on aesthetic perfection. Whether especially rare, consistently popular, or merely parts of the Society's extensive and representative collection of fauna, they were supposed to be species representatives in order to fulfill the Society's principled mission of exhibiting a series of animals from which naturalists and the public might learn about, and take pleasure in, the animal world. Creatures that deviated from the perfection expected of a species representative were often removed from the collection, their value as display objects substantially depleted. Such deviations from physical perfection or normalcy included those with physical deformities or disease and those advanced in age. In 1870, a litter of tiger cubs was born with unspecified deformities, and it was stated that if at all possible these creatures ought to be sold rather than maintained in the collection. The Society clearly sought to cash in some of the animals' latent value, even if the animal commodity was physically damaged. The retention of *some* perceived value was made good through the exchange of these cubs in an 1871 transaction whereby four leopards of various ages and the three deformed tiger cubs were exchanged for a three year-old lioness and £10.[66] Similarly, in 1878, the Society preferred to sell its deformed puma rather than retain it in the collection; it was sent to Liverpool animal dealer William Cross for onward sale.[67] When onward sale was not possible, of course, deformed animals were simply eliminated. When a litter of four lion cubs was born in the Gardens in 1884, three were sold while one, severely deformed, was destroyed. The representative zoological collection was no place for an animal deviant.[68]

Physical imperfections aside, aged and duplicate animals had diminished value as both scientific and commercial exhibits. Some, however, retained value in death through conversion into food for the Zoo's carnivores. As damaged specimens they were sacrificed in order to perpetuate the liveliness of species of greater perceived worth. In 1839, for instance, the "little old" zebu was slaughtered for this purpose, and in 1860, when additional sustenance was required to feed the Society's newly acquired polar bear, two or three animals "not good enough to sell" were surrendered to the cause.[69] Many aging animals could be disposed of because, like diseased or deformed creatures, they were no longer seen as perfect specimens and thus had diminished scientific and commercial value because they could not be deemed representative of their species. Through the wild animal trade, a more aesthetically desirable species representative could be acquired to replace aging or weakened animals currently inhabiting that species' space in the physical catalog of animals. According to the same rationale, duplicates could equally be disposed of because the Society did not really *need* to have multiples of the same species in such a collection.

In all of these ways, the Society's lively commodities were variously, and consistently, valued across the nineteenth and early twentieth centuries according to their capacity to serve the entwined imperatives of science and commerce. However, animals at the Zoo were not only conceived to be one-dimensional items of commercial and scientific worth: they were, in addition, direct *producers* of revenue in and of themselves and not in relation to the way they could stimulate the scientific or visitor gaze. Animals retained their direct financial value as commercial objects after acquisition, and this was frequently translated into cash, or used to cultivate patronage and favor. The commercial aspect of the Society's operation was bolstered through its active reinsertion of animals into the circulatory systems of the animal trading arena. In 1859, it was explicitly stated that "an adverse balance at the end of the year has been avoided by advantageous sales of surplus stock."[70] In 1839, the Society considered selling its rare ungka-puti (agile gibbon) at the same time as it sold its recently acquired chimpanzee for the considerable sum of £150. Justified in light of the Society's poor financial position, the sale of the animal was eased by the promise of a replacement specimen within the year.[71] Likewise, in 1879, the Society's "surplus stock" of lions was sold for £225, and in 1886 £286 of stock was sold to the Continent and Australia.[72]

Just as young animals could draw people to the Zoo, they were also useful as generators of money. Beginning in the mid-nineteenth century, in a culture where exotic pet keeping was becoming increasingly popular, the Society actively attempted to produce young animals for itself through breeding. The 1864 *Annual Report* stated that the General Committee had given attention to the breeding of animals, with the view of having an annual supply of stock and were "endeavoring to breed leopards." They were also on the lookout for a male llama with a view to the breeding and sale of those animals.[73] The extent to which animals were produced through successful breeding is sometimes quite astonishing. In 1894, for instance, it was reported that the Society's lioness Stumptail had no less than fifty-six cubs scattered at various zoological institutions throughout the world.[74]

Selling animals, though, was a rather final means of translating their latent value into hard cash. More sustainably, many zoo animals were loaned out. In the process they were transformed into a kind of renewable resource that could generate income for the Society many times over. In 1840, for instance, the Society capitalized on supposedly having the only ungka-puti in Europe by arranging for its exhibition at the Egyptian Hall in Piccadilly where the public could pay one shilling each to view it. The Egyptian Hall was regularly used as an exhibition space in which people could view interesting objects, be they relics, human unclassifiables, or beasts, at the heart of the Empire. Indeed, Queen Victoria, who had a well-known interest in animals, visited the ungka-puti during its summer on display.[75]

Sales and loans aside, surplus animals were also frequently gifted to individuals and institutions elsewhere, in a system of patronage through which their value was communicated symbolically rather than financially, for the cultivation of connections. In 1838, birds were donated by the Ornithological Society of London, prompting a response from the Society, which promised favors to come.[76] This cooperative system is important. In its incorporation of an array of zoological institutions, it nurtured an environment of collaboration that was to become increasingly pivotal to the acquisition of animals in the midst of a range of seismic political, cultural, and economic shifts over the course of the twentieth century. Thus, it is to these transformations and their implications for human-animal relations rooted in perceptions of value to which we now turn.

WORLD OF WOUNDS

In many important ways, these kinds of estimations of animals as resources with explicit extrinsic values endured unchanged well into the second half of the twentieth century. Accordingly, and like so many zoos all over the world, Bristol Zoo continued to employ methods of acquisition that were astonishingly destructive; animal dealers maintained their position as dominant nodes in the animal trading network; and the transportation of animal commodities from one place to another continued for a long time to be profoundly wasteful. Indeed, the character of the global wild animal trade continued to persist in a form shaped by the synthesis of financial exchange and the cultivation of mutually beneficial relationships. Certainly, donation from sources both at home and abroad remained a vital aspect in keeping the Society's finances healthy: "The Tigress has improved greatly, and is a splendid creature," noted the Society in 1921, "and if we could only obtain an equally good Tiger by purchase or presentation—and preferably the latter—we might reasonably expect, in due time, to display some cubs in our cages . . . let us hope our Indian friends will soon relieve our distress."[77]

Similarly, the acquisition rationale at the core of the Society's activities remained unchanged until well after the Second World War. The collection was still to represent a catalogue of species. Although native species such as foxes, squirrels, and indigenous birds had largely disappeared from the collection by the beginning of the twentieth century, totality remained of vital importance, even as two world wars over the course of four decades, alongside the stringent trading restrictions that came with them, meant that many species became more difficult to acquire than they had generally been during the nineteenth and early twentieth centuries. In 1943, for instance, the Society was able to boast that it had "a large and representative collection," even in light of the difficulties in obtaining animals presented by the economic constraints of wartime.[78]

In conjunction with the scientific imperatives that informed the consideration of animals as equal constituents of a broader whole and that necessitated the ongoing possession of a comprehensive zoological series, popular megafauna, and rarely seen species similarly remained of elevated significance toward the middle of the twentieth century. Just as they had in the nineteenth and early twentieth centuries, rare creatures could satiate the desire to possess animals that were both vital assets to a collection that purported to be scientifically holistic and that could also draw crowds and thus generate revenue that could, in turn, bolster the scientific collection.

The first seventy years of the twentieth century saw the arrival of novelties such as the Society's first gorilla, Alfred. Alfred arrived from the Belgian Congo, firstly from a Greek merchant and then through Rotterdam Zoo in the Netherlands in 1930, at the significant cost of £350 (approximately £11,700 today). He was a creature brimming with charm. Despite having become known to naturalists in the West in the mid-nineteenth century, few living gorillas had been seen in Britain prior to his arrival. For many years, only inanimate fragments of the gorilla, its skin and bones, were available for close inspection. In 1849, five gorilla skulls were acquired by the Bristol Institution (and they remain, off show and locked away, at the Bristol Museum and Art Gallery still) and were believed to be the only specimens known in Europe. These kinds of glimpses of the creature whetted the appetite for a sight of a real living specimen, especially in conjunction with terrifying tales emerging from the African interior.[79] A live gorilla was supposedly displayed as part of Wombwell's Menagerie during the winter of 1855–56, and a gorilla named Mumbo, who arrived amid much media fanfare, lived for just seven weeks at London Zoo in 1887 before he, too, expired. At the end of the nineteenth century, a gorilla at William Cross's Liverpool menagerie was thought to be the only living representative of the species in Europe.[80] In 1930, then, for zoo-goer and zoologist alike, the chance to set eyes upon of a living gorilla was a rare opportunity indeed. The sensation surrounding the display of the so-called Gargantua the Great at a combined Ringling Brothers and Barnum and Bailey show in the United States in the mid-1930s and the popularity of two US zoo gorillas—Bushman at Chicago's Lincoln Park Zoo and Susie at the Cincinnati Zoo—testify to the magnetism of the species as an exhibit at this particular moment.[81] Okapi (1961) and white tigers (1963), both of which were rarely seen in captivity anywhere in the world at the time of acquisition, followed Alfred as the Zoo's headline animals.

The financial values ascribed to the animals in the collection reflect the persistence of the earlier period's hierarchy of biocapital. In 1935, Alfred was commercially valued at a whopping £1,000, and the elephant, Judy, at £700. These were the most expensive creatures in the collection owing not only to rarity—in Alfred's case—but also to popularity with visitors. Charismatic megafauna such as the big cats (£70–150), oryx (£60), and zebra (£50) were the next most expensive, closely followed by the polar bear (£40). By contrast, at the lower end of the market were Java monkeys, badgers, and eagle owls, all of which were valued at just £1 per specimen.[82] Such a hierarchy reflective of judgments of extrinsic worth denotes the ongoing significance of the combination of scientific and commercial interests at the root of the Society's operation and thus its valuation of animal life, as well as the availability of species in the global wild animal trade. All of this continued to color the ways in which animals were framed in a complex value system.

Young animals, too, retained their position as esteemed creatures. They were enduringly popular as exhibits and as a means of generating finance directly through their onward sale. Accordingly, breeding policy continued to reflect that. A female polar bear was specifically requested from London-based animal dealer George Chapman when he offered to procure a twelve-month-old specimen for £75 in 1927.[83] On this occasion, the age and sex of the animal made for an especially favorable lively commodity: an animal ripe for breeding and the production of commercially lucrative cubs. The inclination to reject animals deemed imperfect continued as well, and, indeed, the receipt of damaged animals by the Society occasionally caused considerable anxiety. In 1961 a clouded leopard with a bald patch and infested with parasites was provided by the Dutch dealer Frans Van den Brink. Director Reginald Greed retorted that the animal was "not a good specimen." The animate product was faulty, and Van den Brink consented to supply the Society with a replacement that was more conducive to its designated role as a representative of species.[84]

Despite such unambiguous continuities in the Society's valuation of species, however, the turn of the twentieth century and, later, the decades immediately following the culmination of the Second World War saw the intensification, in quite distinct ways, of a sense that the world was emblazoned with raw wounds, that its soils were stained with the blood of precious wild things that might soon be lost forever. Eventually, this realization reconfigured the layers of a complex valuation system through which wildlife, and animals at the zoo, specifically, had been judged across much of the nineteenth and early to mid-twentieth centuries. Importantly, these attitudes emerged and matured *at the same time* as zoo animals continued to be valued in the ways outlined above. When we trace the history of the emergence of this way of relating to wildlife at Bristol Zoo over the course of the twentieth century, we find the shifting contours of a system of animal valuation that eventually came to position the natural environments of the world as simultaneously worth cherishing and also—still—extrinsically valuable as resources in the contexts of science and commerce. These changes in the philosophical underpinning of human-animal relationships were to have substantial consequences for the Society's physical interactions with animal life, not only at the Zoo but also in habitats all over the world.

At the end of the nineteenth century and in the pre–Second World War period, in line with attitudes relating to wildlife sanctuaries and national parks more widely, zoological societies tended to take the view that species needed to be in danger of extinction *and* considered worthy of protection in order for preservationist action to be undertaken in their name.[85] Those that were deemed to deserve protection appear to have been valued according to somewhat crude criteria that reflected attitudes toward beauty and violence in the natural world. Carnivores tended to be thought of as vicious beasts and thus did not deserve a second thought: a 1936 edition of the

Zoological Society of London's *Zoo* magazine, for instance, clearly positioned American bison as worthy of preservation, while the jaguar was considered a "savage" pest, unworthy of human protection.[86] This kind of valuation structure prevailed for some time at Bristol Zoo. By 1899, a statue of an
Italian shepherd stood at the Zoo's main entrance. It featured the inscription
"protecting the bird from the ravages of a fox."[87] The implication was that the
aesthetically pleasing bird merited protection to the detriment of the hungry
carnivore. Importantly, a sense of *ownership,* an almost inescapable dimen-
sion of human-animal relationships, was another factor that affected species
worth even as a conservationist ethos began to emerge at the Zoo. The 1926
guidebook declared that the raven was fully eligible for human protection
because of its status as a local bird. Its figurative possession not only by the
people of Britain but also of the West of England meant that it should be preserved: "As head of the Corvus family, and an aristocrat of noble lineage and
of great antiquity, the raven deserves the preservation of his race. . . . Much
reduced in numbers, no effort should be spared to ensure his retention on the
British list. . . . In the West of England, as the result of vigilant protection,
he just holds his own. Long may he survive."[88] Similarly, the golden eagle was
revered as a national species. It was a "magnificent bird in every respect, and
one fully deserving preservation . . . its retention as a British species should be
considered an asset to the fauna of the country."[89] In singling out particular
species, exceptional value was assigned to some animals over others for reasons that do not necessarily reflect an *intrinsic* sense of nonhuman value but
that instead reflect human judgments and valuations relating to what constituted the most agreeable characteristics of wildlife. Significantly, too, there
was a fundamental sense that wildlife was especially valuable when it somehow belonged to the people. This anthropocentric system of valuation and
the sustained centrality of human possession clearly cohere with the prevailing attitude toward species across the nineteenth and much of the twentieth
centuries, and it endured into the post–Second World War period. In 1961, a
pair of okapis (a species that is notoriously elusive in its indigenous forested
habitats in West/Central Africa) arrived at the Zoo. In a BBC television program following the arrival of the animals, narrator and marine ornithologist
James Fisher acknowledged that the survival of the species depended on their
breeding in captivity. Director Reginald Greed concurred but instead chose
to emphasize how proud the Society was to possess such a rarity as part of

its collection. The possession of this endangered species lay very much at the heart of its perceived value.[90]

In spite of this centrality, however, after the Second World War ways of understanding the natural world transformed in important ways. These shifts were profound and resulted in transformations in both the way in which the Society valued its animals and in the ways in which it physically related to them as precious beasts. These revolutions reflect both a response to increasing cultural concerns about the state of the planet alongside an allied anxiety about the continued pursuit of exploitative nineteenth- and early to mid-twentieth century collection practices. What eventually emerged at the Zoo between the Second World War and the end of the century was a system of animal valuation that merged a continuing philosophy of human *possession* of the more-than-human world and ongoing commercial demands with a transforming scientific mission rooted in the strengthening ethos and science of conservationism and sustainability.

Consideration of the astonishing destruction of the two world wars, and the ever-present threat posed by the specter of atomic war in the 1950s, 1960s and 1970s, joined with a greater recognition that humanity's industrial and scientific ingenuity was ravaging the wilds of the world. This sense of vulnerability and the imagination of the rest of nature as defenseless victim of human rapacity became encapsulated in the growth of environmentalist movements.[91] Simultaneously, echoes of the Romantic and Transcendentalist veneration of and connection with a wild, beautiful, and terrifying nature lurked within the counterculture of the 1960s.[92] Seismic political developments, such as civil rights and antiwar movements, and protracted decolonization through the demise of European empires also contributed to a cultural milieu in which the exploitation of animals and their environments became increasingly unpalatable. In addition, ways of watching wildlife were fundamentally altered through nature film and, later, television. Films such as Walt Disney's *True-Life Adventures* series (1948–60) and *Born Free* (1966) and productions from the new BBC Natural History Unit, which was itself based in Bristol, invigorated an increasingly sensitive and appreciative attitude not only toward the rest of the natural world as it supposedly existed beyond the curious gaze of people but also toward institutions whose structures seemed all the more oppressive in opposition to the Edenic environments beamed in ever greater color and clarity into the nation's living rooms.[93] Peter Singer's highly influential 1975 *Animal Liberation* channeled some of these senti-ments into an appeal for animal welfare and the liberation of beleaguered beasts from institutions like farms, laboratories, and, of course, zoos.[94]

These cultural revolutions transformed the ways in which many people understood the role of zoos and their relationships to the nonhuman crea-tures living both within and beyond their walls. They are important changes indeed. They underpin so many of the changes in human-animal relation-ships across this book, and so we will return to them again and again as we descend through the layers of interspecies relationship that we encounter at the zoo. These sociocultural instabilities were not enough on their own, though, to stimulate transformations in human-animal relationships in these kinds of captive spaces. The economic impact of the Second World War and the introduction of legislation designed to regulate the transnational move-ments of endangered species meant that acquiring animals through the trad-ing channels that had remained reliable over the course of the nineteenth century and much of the first half of the twentieth century became increas-ingly difficult. In 1937, for instance, the Society met difficulties in arranging for the shipping of king penguins from South Georgia. Protracted correspon-dence between the Society, the Ministry for Agriculture and Fisheries, and the colony itself concerning the procurement of the appropriate licenses and permits for import delayed successful acquisition well into 1938.[95] Of even greater significance insofar as the Zoo's continued operation in the global wild animal trading arena was concerned was the 1973 formation of the Con-vention on International Trade in Endangered Species of Wild Fauna and Flora (CITES), involving 175 international parties. The legislation imposed a raft of restrictions, not least on the trade in many species of threatened animals. These changes, coupled with increasingly stringent quarantine reg-ulations, restricted and constricted the mechanisms of the traditional above-ground networks of animal acquisition within which the Society had previ-ously operated, and this meant that human-animal relationships could not manifest in the ways that they had previously done so.

All of this meant that zoos were forced into a corner. They needed cer-tain animals in order to make money and thus stay open. In order to ac-quire the kinds of species that were understood to be commercially valu-able in terms of drawing in the crowds, they had to start working formally with each other on a global scale in order to move biocapital among themselves. Of course, and as discussed, they had worked together throughout the nineteenth and twentieth centuries, but they had largely done so within the limits of a vast network of casual correspondence. In order to establish a breeding body of animals that could reliably supply valuable display specimens, zoological collections organized themselves into official collectives that operated on a variety of geographical scales.[96] Two of the first such collectives were the International Union of the Directors of Zoological Gardens (IUDZG), founded in 1935, and the Federation of Zoological Collections (FZC), formed in 1945. Through these organizations, zoos formally assisted one another in the maintenance of stock that represented both scientific and commercial value through reciprocal exchange. Such organizations set the stage for the emergence of further collaborative organizations such as BIAZA (the British and Irish Association of Zoos and Aquariums, founded in 1966), EAZA (the European Association of Zoos and Aquariums, founded in 1992), and WAZA (the World Association of Zoos and Aquariums, which grew out of the IUDZG), all of which endure today.

Consequently, the establishment of formal cooperative breeding schemes whereby zoos loaned individual animals to each other for the purposes of producing biocapital for themselves emerged out of these kinds of organized collectives in the 1960s. Early exchanges were, however, very much rooted in the appropriative attitudes toward nature of the pre–Second World War period. Undoubtedly principally motivated by a desire to generate possessions rather than to advance conservation agendas, discourse surrounding them explicitly used the vocabulary of ownership.[97] The first exchange concerning Bristol Zoo took place in 1968 when the female polar bear Sabrina was mated with the Zoological Society of London's Pipaluk. There was no mention of any conservation value underpinning, or even peripheral to, this exchange. Similarly, a 1975 cooperative breeding scheme involving London Zoo's female gorilla Lomie explicitly referred to the animals concerned as "property."[98]

As if these changes were not enough, the fortunes of UK zoos steadily declined from their heyday in the 1960s. By the mid-1980s the financial difficulties in which many UK zoos found themselves reflected a rising distaste, emerging because of the shifts in the post–Second World War cultural environment as explained already, for the maintenance of wild creatures in captivity.[99] In isolation, cooperative breeding was therefore pointless in the context of falling visitor numbers. In the midst of these powerful changes, then, zoos were obliged to reinvent themselves from top to bottom as *essential* refuges for an innately valuable, fragile, and threatened wild world.

At least, that was the new *public* face of the zoo.[100] Lurking behind the facade that was erected from the mid-1980s were massively varying levels of commitment to the conservation cause, not only among zoological societies themselves, but also over time and according to context. Consequently, zoos have been accused of greenwashing their operations, of exaggerating the extent of their conservation roles in order to justify their continued existence.[101] At Bristol Zoo, while there was a strong desire to work toward saving species, there remained the very same commercial pressures that lay at the heart of the nineteenth-century zoo. Zoo animals toward the end of the twentieth century did not, then, simply transform into beasts regarded as of intrinsic rather than extrinsic value. Their layers of value simply realigned in kaleidoscopic fashion in response both to cultural and economic changes and to the perception of an increasingly wounded world.

THE VALUE OF BIODIVERSiTY

The 1980s and 1990s were transformative decades. Animal values and human-animal relationships at the Zoo were reconfigured and realigned. The process of change was complicated and was often a source of internal conflict and contradiction during which enduring commercial and newly emerging scientific values jostled for prime position. At the beginning of the 1980s the Society was active in "a concerted international effort to safeguard" the okapi. When an individual was born in the breeding herd at the Society's 136-acre Hollywood Tower Estate just north of Bristol in 1983, it was hoped that he could be exported to Europe in exchange for "blood"—a euphemism for healthy genetic material—emanating from herds held there. In so doing, the Society would be promoting the conservation of a disappearing species. Through collaborative action, an okapi "blood pool" was being established; the bodies of the captive creatures concerned served as vessels whose physical translocation allowed for the movement and manipulation of a prized bloodline, of a precious genetic cargo.[102] In the very same year, however, some members of the Society were anxious about the use of the term "conservation" itself. It was problematic, they worried, not least because it seemed to necessitate expensive work, and this was something to which they were reluctant to commit.[103]

By 1991, the Society had realized the importance of publicly portraying itself as an active steward of a vulnerable natural world and was consciously attempting to "be seen" taking part in local conservation initiatives.[104] The vision of the world that had emerged was one in which humans and wildlife existed in an entangled relationship with one another. In order to advance the new mission that reflected this conceptualization and associated value system, the Society outlined its new central undertaking "to maintain and defend biodiversity through breeding endangered species, conserving threatened species and habitats, and promoting a wider understanding of the natural world."[105] All of this appeared to recognize an *intrinsic* value in non-human animal life. In the process, it placed nonfinancial, and non-extrinsic, values on all species across the entirety of a holistically considered natural world. This kind of value system was enshrined in the iconography of the "stationary ark," which appeared alongside the Society's new mission statement from the early 1990s and which became popular among many of the world's zoos.[106]

If we take this kind of valuation of animal life on its own for a moment, it is clear that this apparently simplistic estimation itself represents a multilayered system of animal estimation. In many contemporary zoos, it has been suggested, it is the gene and not the individual that is considered to be of most value.[107] In fact, the relationship between the various components of the more-than-human world with which many zoos have become concerned — gene, individual, ecosystem, biosphere — are markedly more complex than this allows. It is true that within this system of conservation-based valuation the genes of endangered animals are cherished as never before. Though they hold the capacity to facilitate the survival of species, they are nevertheless, in the context of most zoos and for now at least, redundant without animal bodies to act as vectors for their transmission.[108] In turn, healthy genes and healthy bodies are considered vital to the perpetuation and re-creation of healthy ecosystems that are essential to sustaining species once they have been rescued from the brink.

Consequently, the coordinated exchange of genetic material — one symptom of this transformed value system — has led to the establishment of transnational networks of zoos and animals that are no longer predominantly buttressed by global circulations of finance. Species studbooks and Zoological Information Management System (ZIMS) reports tell us much about the complexions of many of these networks and the human-animal relationships at the Zoo. In 2011, for instance, a male red panda, Sir Ed, was flown 19,000 kilometers from the Wellington Zoo in New Zealand to engage in a captive breeding program with Bristol's female red panda, Jasmina.[109] Using an information system called the Single Population and Records Keeping System (SPARKS), a combination of genetic material, and thus a pairing of individuals, which avoided significant in-breeding, was arranged. In this way, healthy offspring could be *created* that would proceed to contribute to an ever-stronger gene pool, which would serve the purpose of aiding efforts to maintain the global captive population of this species. For instance, an Asiatic lioness that departed the Zoo in 2014 has genetic material deep within her flesh and bones that ties her to lions still in Bristol in 2017, as well as to others in Lodz, Helsinki, Edinburgh, and Junagadh.[110] When her two most recent—male—cubs eventually move to other zoos as part of the Asiatic lion conservation breeding program, the genetic network will expand once more. In so doing, it will perpetuate the formation of a complex and (for now) expanding transnational genetic network. These kinds of operations are extensive in today's Zoo, with the Society presently involved in over one hundred such programs.[111]

Unlike earlier physical manifestations of a longing for animals, this is not about the overt mastery of extraction and possession of individuals but rather a stewardship rooted in a more benign mentality of care, human ingenuity, and a mastery of destructive, if anthropogenic, forces. This is, according to Irus Braverman, a manifestation of power, of the control through care of a delicate world.[112] Yet, despite outward appearances, and as Braverman recognizes, such activities and approaches remain both the means and products of manipulation. Just like the wild animal trade of the nineteenth and twentieth centuries, animal movements in a gene-oriented conservation system are intensively managed. The conservation of species is, at its heart, a manipulative act that relies on the ability of ingenious and compassionate human animals to capture, confine, and control the bodies and biologies of their captive wards.

The twenty-first-century Zoo today houses a predominantly captive-bred collection principally established through this noncommercial movement of animals between zoos. Improving the health of an *entire* captive species is paramount to the Society's broad mission, and that itself has shifted the ways in which particular captive species and individuals have been valued. Today, if a specific captive animal presents a significant physical abnormality, then its ability to continue as part of the collection is considered according to its capacity for suffering as well as its importance in captive breeding programs. If it was physically able to reproduce and was part of a species considered to be of critical conservation significance, it would not be destroyed, as it would have been in earlier years, but, rather, would be maintained in the collection and bred.[113] This is not about the physically deviant animal now having a home at the zoo. Rather, it is about the intrinsic worth of the imperfect individual *in conjunction with* its precious blood. Similarly, where high production of animal offspring once sent the pound signs spinning, overproduction and unchecked reproduction within this transformed valuation system is generally considered a hazard to species survival. Uncontrolled breeding is damaging, for it might lead to the propagation of inbred captive populations with narrow and weak gene pools that lack broad and deep genetic variety. This is a significant risk given that the whole point of captive breeding programs is to create viable—healthy—populations. A shallow gene pool contains all sorts of dangers; not least, the possibility of the perpetuation of genetic disease, as was the case with the Society's famed white tigers between the 1960s and 1980s. Contraceptives were introduced from the 1990s in order to control the breeding of a number of species likely to perpetually reproduce, such as Persian leopards and Sumatran tigers, both of which were involved in international breeding programs. In contrast to practices elsewhere (such as Copenhagen Zoo's now infamous decision to euthanize Marius the giraffe—as part of a supposedly widespread "zoothanasia" campaign—in order to eliminate his genes because they were "too common"), contraceptives continue to be deployed at Bristol Zoo as a means of managing precious genetic lineages and thus maintaining their high value.[114]

Increasingly integral to these transformed conservation imperatives, and commensurate with the aspirations of the International Union for Conservation of Nature (IUCN), was an increased impetus to return captive animals to their habitats and to work in threatened environments to both enrich the health of ecosystems and promote a sustainable future for both people and wildlife.[115] While the movement of animals from captive to wild environments represents something of an inversion of the destructive acquisition processes of the nineteenth- and twentieth-century wild animal trade, these operations are also symptomatic of transformed human-animal relationships more broadly. They show that in the conservation age it is not only the gene and the physical body that have been attributed intrinsic value. Rather, the gene and the individual animal body exist as part of a matrix of significance that also includes habitats, ecosystems and, most critically of all, the very foundations of human interactions with wildlife in every part of the planet.

In 1994, the Society contributed to a managed reintroduction scheme for the first time, collaborating in work with Polynesian tree snails on the island of Moorea in French Polynesia. This effort was unsuccessful, and Bristol Zoo is now the only place in the world where these creatures, extinct in the wild, can be found.[116] In 1999, the Society achieved its first independently managed reintroduction when it released endangered Barberry carpet moths back into their native habitats in Gloucestershire, and in 2003, water voles were successfully reintroduced into habitats in Somerset.[117] Good intentions notwithstanding, these instances, centered mainly on native species, presently stand in isolation. The re-wilding of captive species as a material expression of the intrinsic valuation of species, ecosystems, and sustainable human-nonhuman relationships remains an ideal rather than an actuality. Internationally, the process is still in its infancy; examples of successful reintroduction are minimal, and much skepticism remains, even among zoo and conservation professionals, about its long-term viability.[118] Perhaps that is not the point. The green shoots of genuine action that these activities represent nevertheless denote a transformed conceptualization of wildlife; one rooted more than ever before in benevolence and respect; a metabolism coming back into balance.

Again, reflecting this kind of ethos, work in situ (i.e., in indigenous environments and arguably neocolonial in character) has largely focused on educating local peoples and encouraging practices designed to promote sustainable relationships with species and their ecosystems. Local people in places as far apart as Cameroon and Clifton are, through the Society's work beyond its boundaries of brick and iron, educated to live as interwoven entities in a world of both human and more-than-human making. The integrated nature of Bristol Zoo's conservation agenda in a "One Plan" approach, which includes cooperation between zoos and other organizations and the concurrent exchange of knowledge, illustrates its intention to contribute effectively to conservation initiatives on a deep, global scale.[119] One such project at the Dja Biosphere Reserve in Cameroon, in collaboration with Ape Action Africa, entails the instruction of local people so that they might live in sustainable relationship with their wider environment. Unsustainable practices at the heart of indigenous lifestyles, such as deforestation and the trade in bush meat, are all challenged. Closer to home, the ongoing Avon Gorge and Downs Wildlife Project has educated the public about sustainable living since 1993. At the same time, it works to conserve rare plants and animals in the area, including the Bristol and Willmott's whistlebeams (species found nowhere else in the world), peregrine falcons, and several species of rare wildflower.[120]

The emergence of this conservation-oriented valuation system, combined with changing international trade regulations and a mounting distaste for animal exploitation, brought about important revolutions in stock policy at the end of the twentieth century. Increasingly, it was realized that the maintenance of a comprehensive, representative collection was neither possible nor desirable. For many of the same reasons, stock policy also departed from the privileging of the maintenance of the more prosaic zoo species, as it did in many of the world's urban zoos.[121] In 1985 the Animal Committee reported that "Bristol Zoo, like many of its sister establishments, is considered a 'mainstream' zoo with a wide-ranging, representative collection." It continued: "the modern trend and demands . . . necessitate that virtually all zoos . . . cannot keep all species." This appreciation of changing expectations marked the beginning of the gradual dismantling of the Zoo's traditional collection just as the multilayered conservation ethos described above was beginning to emerge.

The fall of one policy and the rise of another in its place *appear* to fit seamlessly together. Rhinoceroses, difficult to procure and house in captivity, made way for the smaller pygmy hippos, a species that had been identified as critically endangered and that had just as much charismatic value. The ape collection, an exhibit previously comprising a number of different species, was broken up in the mid-1990s, and gorillas were exhibited as representative of anthropoid apes as a broad taxonomic family. Like the Asiatic lions discussed at the beginning of the chapter, gorillas were transformed into flagship zoo species, charismatic species onto which conservation messages can be focused.[122] In place of many of the larger mammals, smaller, threatened species such as insects, reptiles, and amphibians, most of which had significant conservation value, were acquired in greater numbers through captive breeding programs. This transition was actually far from seamless, however: while the scientific rationale underpinning the Society's activities may have transformed, the Zoo, like so many others across the country, remained an independent institution that was not only reliant on the generation of income but also just one among a growing multitude of tourist activities emerging in the later decades of the twentieth century. Such commercial pressures meant that a stock policy more clearly aligned with an emerging conservation ethos was complicated by the ongoing need to exhibit animals that the public valued and would pay to encounter.[123]

Star animals remained of great importance even in the midst of the recon-figured human-animal relationships of the late twentieth-century zoo. The continuing provision of a spectacle that met the expectations of a visiting public who had largely been brought up visiting zoos stocked to the brim with favorites like the elephant, polar bear, lion, and tiger seemed to sit at odds, not only with the emerging conservation ethos but also in the context of a heightened cultural sensitivity to exploitation. In 1990 the Society began to recognize that the continued exhibition of elephants would leave it open to charges of animal abuse in this kind of cultural climate.[124] Yet elephants re-mained, to many, *the* zoo animal. The solution, they proposed, was the estab-lishment of a rolling program in which the Zoo would house baby elephants. When they matured to a size and age where they were unmanageable, they would be moved on. While the idea was marred by concerns relating to the feasibility of import and severe misgivings that the international community of zoos would be in any position to unendingly receive large numbers of adult elephants in the future, the plan nevertheless illustrates the perpetuation of an extrinsic judgment of animal value, even in the age of conservation. Such was the strength of feeling surrounding the need to continue the exhibition of quintessential charismatic megafauna, however, that an African plains ex-hibit was planned in 1991. The idea was that it would display African animals in a large space spanning the lake, itself a central feature of the Zoo Gardens. In effect, the exhibit would dominate the entire zooscape, significantly alter-ing a spatial character that had endured, in some ways, for more than a cen-tury and half. This scheme did not reach fruition, either.[125] Three years later, in 1994, some members of the Council suggested that the Society should seriously consider continuing the display of the stereotypical zoo megafauna, even if their conservation value was low. The solution, of course, was that emphasis should be placed more squarely on the acquisition, maintenance, and display of *large endangered species,* effortlessly combining commercial and conservation imperatives, extrinsic and intrinsic valuations.[126] Soon after the death of the last white tiger, the Society began displaying the popular *and* endangered Sumatran tigers in their place.[127]

Young versions of popular animals have likewise remained especially valuable commercial propositions for the Society and offer the chance to combine various valuations of nature. The production of offspring through coordinated captive breeding programs might officially be presented as integral to the saving of species, but these young animals also continue to be employed as a means of generating press interest and, by extension, visitors and revenue. The media interest surrounding the 2011 birth of a young gorilla, Kukeña, and the 2011 announcement of the birth of twin lion cubs, Kamran and Ketan, are testament to the value of young animals to the Society's irrevocably entangled conservation and commercial operations.[128]

In addition, rarities retained their commercial significance at Bristol. During the Christmas season, beginning in 2008, the Zoo has temporarily displayed reindeer (usually as part of a larger Santa's Grotto scenario) and has energetically promoted their "flying" visit to the Gardens.[129] The creatures have been positioned, for a limited time only, alongside the meerkat, lion, gorilla, and lemur as among the central sights of the Gardens. While they are far from rare in terms of free-living populations, their exaggerated prominence at Christmas relative to other times of the year makes them *appear* to be so; they become exoticized, if only for a short period.

The 2011 opening of Meerkat Lookout, in particular, speaks volumes about an enduring commercial agenda at the heart of the Zoo. The display of the animals reflects the present popularity of these charismatic (mini)fauna. Films such as Disney's *The Lion King* (1994), the Discovery Channel's *Meerkat Manor* (2005–8), and a recent spate of advertisements for a price comparison website have brought the animals firmly into British popular consciousness.[130] Classified as "of least concern" by the International Union for Conservation of Nature, the display of these creatures illustrates the continued extrinsic valuation of animal life in light of their ability to attract paying visitors seeking amusement. In combination, then, the various extrinsic and intrinsic values assigned to captive animals present a perplexing picture of layers within layers of human-animal relationships in the twenty-first-century zoo.

CONCLUSION

Over the course of nearly two centuries, the values attached to animal life according to the objectifying and homogenizing imperatives of science and commerce have remained a foundational layer of human-animal relationship, and one often framed in a global context. Through the consideration of animals as precious objects of desire in both transforming and enduring ways over the entire period, their bodies were consistently subjected to a range of manipulations. Across the course of the nineteenth and early twentieth centuries, these systems of valuation and the manipulations at their heart took the form of extraction on a gigantic scale, followed by translocation across vast distances, and these enterprises were of an enormously wasteful character. However, over the course of the twentieth century the nature of this foundational relationship began to transform, or rather it became increasingly complicated as its contours reconfigured. In a world threatened by human activities on a global scale, value was increasingly attached to genetic material, with precious animal bodies serving as vessels transporting that information from place to place. These bodies and the genes within them were reconceptualized as servants of the preservation of species and ecosystems. Yet this system of valuation ran parallel—indeed it was entangled with—the continuing extrinsic valuation of animals as objects of science and as objects of commerce. Thus, at the Society's inception in 1835 and in the first decades of the twenty-first century transformed human-animal relationships rooted in value judgments in this captive world were forged in the same frame: the competing/complementary cultures of science and commerce.

This layer of human-animal relationship was consistently unstable and hierarchical. Both extrinsic and intrinsic values attached to animals rested on species, sex, age, and condition among many other factors reflecting a deeply rooted hierarchical speciesism at the heart, not only of zoo culture, but also of human-animal social lives themselves. What is more, this layer of value reflects a further foundational relationship: the supposed separation of humans from animals and of nature from culture. In the nineteenth and early twentieth centuries, animals were objects to be *owned* by humans, and by the early decades of the twenty-first century animals were entities not only to be owned but also to be *saved* by the ingenious and masterful human being. Even in the conservation age, these are not human-animal relationships defined by equality. Today, as it was nearly two centuries ago, valuations of animals and their natural worlds rested on their relationship to anthropocentric concerns.

2 🐒

The Theaters of Animality

Imagine arriving at the Bristol Zoo in the early 1840s, only a few years after the site had been opened to the public in the summer of 1836. After passing through the entrance lodge, turning left and then walking past the pheasant aviary, the first large exhibit to be encountered on the Main Terrace—the Zoo's central artery—was the Terrace Menagerie. A series of three connected houses, individual museum-style cabinets that were typical in zoos of the period, displayed monkeys, baboons, genet cat, marmot, opossum, big cats, hyena, porcupine, vultures, an array of ruminants, wombat, cassowary, and more. Small and robust, stark iron bars physically separated animals and visitors from each other. The compartments clearly demarcated the animals according to their species so that their specific adaptations could be examined in relation to each other. Here, too, the Serpent Box displayed boas and rock snakes behind a glass screen, for iron bars were obviously inadequate means of ensuring snakes remained securely under lock and key. Proceeding eastward along the Terrace, an aviary for small gallinaceous birds was followed by the Chimpanzee House. At the end of the Terrace stood a structure that echoed the animal amusements of an earlier age: the Bear Pit, a common feature of early zoos, contained "a pole . . . erected for the purpose of showing with what readiness the animals can climb, which they are generally willing to do, if the visitor will only tempt them with a piece of bread or biscuit" (fig. 5).

Russian Bear
"Jack"
Clifton Zoo.
1641.

FIGURE 5. Russian bear Jack on the bear pole, c. 1916. Viner & Co.
Publishers. (Roy Vaughan Collection, Bristol Zoological Society Archive,
courtesy of the Bristol Zoological Society)

To the right, as you venture beyond the pit, were kennels for the Society's collection of dogs, a pond for the otter, and an enclosed portion of the lake for gulls and ducks. Continuing around the Zoo's perimeter, further compartments included those for raccoons, wading birds, squirrels, parrots, fancy pigeons, emu, and kangaroos. By the time you reemerge at the western entrance lodges, you have looked at a highly ordered unnatural kingdom, in which animals are categorized according to kind, whether that be species or class, and in which their forms and behaviors are exposed for examination and amusement.[1]

When the Society's inaugural collection of beasts was officially unveiled in the summer of 1836, it was the first permanent scientific display of living exotic animals in Bristol. Before then, cockfighting, bear baiting, equestrian shows, and even a "sapient pig" named Toby, who was supposedly able to read, spell, and read minds (he was displayed at the annual St. James' Fair held in and around the twelfth-century churchyard of St. James'), were the sum of the city's eclectic animal entertainments.[2] These amusements were just that: innocuous pleasures designed to distract. The animals at their core were conceived of as objects existing for people in pursuit of enjoyment. In somewhat similar fashion, once the Society had acquired precious beasts for its debut collection in the mid-1830s, their objectification was fortified through exhibition practices that converted them into entities of multifaceted encounter. Over the course of much of the nineteenth and twentieth centuries, zoo creatures were exhibited in ways that emphasized the instrumental values assigned to their fascinating, terrifying, sometimes hilarious bodies. The bear pit, for instance, was a structure designed to summon the delight wild things could muster, while simultaneously showcasing the forms and natural behaviors of the creatures in a free-living state.[3]

Converting animals into objects of entertainment is a consistent feature across the entire history of the Zoo. Every animal put on display was definitively *to be looked at,* carefully positioned as if on a stage, performing their originality and their animality for the Zoo's patrons. Yet, the ways in which animal life was staged transformed over the period and especially over the course of the final decades of the twentieth century when, as discussed, the Zoo underwent a revolution in both its public and institutional personas. Changing public sensibilities and evolving technologies combined to bring about important shifts in the ways in which animals were perceived and, sub-

FIGURE 6. Seal and Penguin Coasts, c. 2000. (Bristol Zoological Society Archive, courtesy of the Bristol Zoological Society)

sequently, in the character of the zoo's animal attractions themselves. Thus, just after the turn of the twenty-first century a significantly altered zooscape met the eyes of curious visitors. In place of cramped and bare compartments separating animals according to species, the Terrace displays consisted of animals integrated into theatrical, barless arrangements informed by habitat. Twilight World (1996) presents animals that live primarily in the dusk and dark, while the World of Water Aquarium (1984), situated on and incorporating the former Bear Pit, exhibits marine life in an array of aquatic environments. A number of immersive experiences such as Seal and Penguin Coasts (constructed at a cost of about £2.8 million in 1999) (figs. 6 and 7), meanwhile, invite visitors to travel *through* theatrical captive spaces, breaking down the barriers between observer and observed. This reflects human-animal relationships that are, in some important ways, distinct from those that permeated the fabric of the Zoo over the course of much of its life.

The natures of the stages on which animals *perform* have transformed in important ways. Visitors have been integrated more tightly into the zoo show. They are able to touch the naturalistic surfaces of the displays and experience proximity to animals and an exotic world that *appears* more authentic and less

FIGURE 7. Seal from within the underwater tunnel, c. 2000.
(Bristol Zoological Society Archive, courtesy of Andre Pattenden)

explicitly confined than ever before. Glass, moats, and near-invisible zoo mesh have replaced thick iron bars as methods of constructing an animal performance based on the illusion of freedom.[4]

Glass enclosures, into which visitors *look,* sit alongside the highly theatrical Seal and Penguin Coasts through which visitors *move.* Indeed, at any single point across the entire span of the Zoo's history, the zooscape was multilayered. In this context, interspecies relationships have been framed by the physical zoo world itself: the very contours of the site of captivity. In this way, there was a diversity of staged exhibits and encounters to be had, and the changing composition of these performances, not only in terms of how displays were configured but also in how visitors were able to interact with animals, reflects an assortment of simultaneous social constructions of animal life at any single point in time. Yet, while the artificial worlds of captive creatures changed in important ways during the Zoo's history, there remain some striking continuities that denote the presence of a deeply rooted objectification and instrumentalization of animals in light of their malleability as objects of entertaining and educational encounter.[5]

Putting animals on the literal and metaphorical stage constitutes a second important layer of human-animal relationship in this captive context and beyond. In our own age, television programs and nature films do exactly that, while YouTube channels devoted to what are presented as the secret, hysterical, and adorable lives of animals—especially cats—testify to our continuing interest in the ways in which animals might entertain and inform us. In many ways, we immerse ourselves in a theater of animality.

CRAFTING THE ZOOSCAPE, 1836–1980

Hardheaded containment of animal life was at the heart of zoo display over the course of the entire period. Zoos are, in whatever form they take, *captive* spaces, designed to amass, *contain,* and display animal life. The initial layer of the construction of the animal spectacular, then, was the apparatus of containment. The zoo, after all, is a symbol and manifestation of mastery.[6] During the nineteenth century and, in many cases, well into the twentieth century, strong iron bars ensured that dangerous animals of both the human and nonhuman variety were unable to jeopardize each other's well-being in circumstances where alternative barriers like glass were too weak to contain them. From the outset, most animals had their own small compartments "faced with strong iron bars" while many of the display areas also incorporated a long iron railing along the front of the display that kept the public at a safe distance from the animals.[7] Strong glass panels later, and more frequently from the middle of the twentieth century, served the same purpose. Other display spaces over the course of the period, including the Bear Pit, the Monkey Temple (1928), the 1935 polar bear enclosure, and the rhinoceros paddock (1953) (fig. 8) made use of pits rather than bars or glass for the purpose of maintaining the state of captivity.

FIGURE 8. Rhinoceros Paddock, c. 1954. (Bristol Zoological Society Archive, courtesy of Drake Educational Associates)

Designing animal spaces was, however, as much about effectively containing living things as it was about communicating ideas about humans, animals, and their respective places in the universe. At the heart of the construction of the Zoo's exhibitions lay various logics about how animals should be grouped, and these reflected the Society's educational imperatives in varying ways. As we saw during our visit to the Zoo in the 1840s, at the outset of and well into the twentieth century species were usually presented individually in small, bare compartments that decisively positioned them as subjects of an intensely curious gaze (fig. 9).

The Society's core educational undertaking and the instrumentalization of animals as objects of instruction at its heart were clearly embodied in this kind of display practice. To a significant degree, scientific logic rooted in Linnaean taxonomy lay among the philosophical foundations of this type of exhibition practice. Through the separate display of animals in bare cages, the attention of visitors was directly focused upon the anatomical distinctions between species and among classes. This reflected the cultural supremacy of a particular scientific rationale that reified the supposed order of nature

FIGURE 9. Lion cage with crowds, c. 1905. (Roy Vaughan Collection, Bristol Zoological Society Archive, courtesy of the Bristol Zoological Society)

inherent in morphologies. In 1840, for instance, two partitions of brickwork were erected in the Quadruped House for "the purpose of making a better classification of the Animals."[8] Displayed as examples of four-footed animals, and as representatives of particular species, these creatures were positioned as specimens inhabiting a particular place in a highly ordered though nonetheless complex world. The collection of the 1840s was primarily arranged according to this taxonomic system, and commensurate with a largely static scientific and educational founding principle, this organizing rationale endured at Bristol wholly uncontested until the mid-twentieth century.[9]

Although the Society positioned its animals within such a scientific schema, the extent to which the public perceived them as such is more difficult to ascertain. Nevertheless, some visitors did gaze upon animals as objects of enlightenment. Schoolboy Frederick George Melville related in 1879 that "the study of Natural History is at all times an interesting subject, to both old and young." He continued by recounting his visit according to the order by which he encountered the animals on display. Florence Grinfield, author of the unofficial guide to the Zoo, also remarked in 1894 that "it must be felt by parents to be a great educational advantage for their sons and daughters, to have such a zoological collection in their midst," through which they had the benefit of comparing "what they have read in their books with the living specimens before them." Another visitor requested a copy of a guidebook to be sent to him in advance of his visit in 1911 so that he might devise the most appropriate inspection of the collection.[10]

Human-animal relationships founded on these kinds of educational encounters also, however, did not rest on the observation of morphologies alone. Sporadically, and increasingly frequently after the turn of the twentieth century, animals were simultaneously contextualized in relation to their habitats. Just as they could in other spaces that metaphorically shrank the world, such as the British Museum and the Royal Botanic Gardens at Kew, visitors to zoos were increasingly able to vicariously travel the globe in only a few hours.[11] At Bristol Zoo, as elsewhere, species were often presented in the context of exotic worlds and, occasionally, within the context of the British Empire itself.[12] In this way, the animal form served as a window to another world, and at the same time, those exotic worlds enriched the captivating colors of the beasts.[13] For a long time, however, this association remained confined to exhibit signage and, more importantly, in the form of official guidebooks. Nineteenth- and early twentieth-century guidebooks were designed to accompany the visitor as they inspected the animals on display. Often, these publications explicitly, and rather anthropocentrically, emphasized the provenance of the animals in the collection with some, though by no means exclusive, focus on their relationship to Imperial territories and activities. The 1843 guidebook related attitudes toward animals among the Brahmins and "Hindoos," colonist engagement with emu and ostrich in Australia, as well as animal superstitions among indigenous Americans. The guidebook appearing in 1882 likewise referenced human-leopard encounters in Ceylon, lion trapping in the Gambia, encounters with crocodiles in India, and American black bear hunting in North America. Although this anthropocentric practice had significantly lessened by the end of the nineteenth century, guidebooks nevertheless continued to communicate a strong sense of connection between the species in their display compartments and the places from whence they came.[14]

In similar fashion, specific forms of naturalistic display—displays that were supposed to mimic natural environments—emerged in the early part of the twentieth century (and, incidentally, some thirty years after their

FIGURE 10. The Monkey Temple, c. 1929. Hepworth. (Roy Vaughan Collection, Bristol Zoological Society Archive, courtesy of the Bristol Zoological Society)

introduction in the wildlife galleries at Bristol's civic museum), which rendered this association between species and place increasingly physical. The Monkey Temple, for instance, was erected in 1928. Based on the Cold Lairs of the Bandarlog (the monkey people's fortress) from Rudyard Kipling's 1894 *The Jungle Book* and resembling a similar 1881 display at Manchester's Belle Vue Zoo, the display echoed the architecture of the Indian subcontinent. It placed the monkeys, which were usually rhesus macaques, within an environment that rendered explicit the connection between the species on display and their illusory place of origin (fig. 10).[15]

Later, the supposedly pioneering volcanic coasts penguin exhibit (1951) was constructed to replace the Victorian cage in which the semi-aquatic birds had previously been exhibited. Naturalistic in style, the enclosure presented the creatures in a way that evoked a scene one might observe from the deck of an ocean vessel on the southern seas (fig. 11).[16]

Some other animals could also be contextualized in relation to their indigenous environments beyond the bounds of their enclosures. The elephants Judy and Rosie, at least, were often adorned in ways suggestive of Eastern splendor, positioning the elephant ride, and the elephant itself, as an attraction within an exotic though anthropocentric frame. In the process, a

FIGURE 11. Penguin volcanic coast exhibit, c. 1960. Harvey Barton & Sons Ltd. (Roy Vaughan Collection, Bristol Zoological Society Archive, courtesy of the Bristol Zoological Society)

sense of the imperial hunting and ceremonial uses of elephants in the British Raj was deliberately evoked in order to deepen the colors of the creatures on display (fig. 12).[17]

While firmly rooted in the science of Linnaean taxonomy, an increasingly physical manifestation of the connection between the Zoo's animals and their place of origin became increasingly apparent over the course of the late nineteenth and early-to-mid-twentieth centuries. This is important, for by the time zoos appeared, and certainly over the course of the nineteenth century, seeing animals was no longer enough on its own because people had become habituated to seeing exotic animals on their doorstep, even if only fleetingly in the form of traveling menageries, for instance. Instead, zoos needed to evoke both wild things *and* their wild worlds, and after the middle of the twentieth century, this system of exhibition was integrated into a new way of grouping species. From the 1950s, an increasing number of displays were informed not only by Linnaean classification but also by species groupings according to zoogeographic distribution. In so doing, the performance of wild *place* was enhanced and the ability of captive animals to serve as portals to other worlds maximized.

In 1954 the world's first nocturnal house was designed by Dr. Richard

FIGURE 12. Rosie with keeper Tom Bartlett, c. 1950. Harvey Barton & Sons Ltd. (Roy Vaughan Collection, Bristol Zoological Society Archive, courtesy of the Bristol Zoological Society)

Clarke and opened at the Zoo so as to present animals not in the primary context of their morphological relationship to each other but, rather, in relation to their shared habitat of darkness. The appearance of the nocturnal house as a habitat exhibit at this point was an aberration; no similar displays appeared until the final quarter of the twentieth century because the imitation of nature at the core of these exhibits was both costly and technologically complicated. Excluding light in a nocturnal exhibit was relatively easy compared to the re-creation of steamy rainforests or baking savannah. The Nocturnal House itself did not last (though we shall make a repeat visit shortly). It was made into an enclosure for gibbons in 1968.[18]

Regardless of the didactic messages the Society wished to convey about the animals and their worlds it was not able to wholly dictate the ways in which the public consumed its spectacular beasts. The purposes of the Society and the changing sensibilities of the Zoo's visitors often interacted in the production of a transforming instrumentalization of animals through exhibition. Consistently intrinsic to the display of animals over the course of much of the nineteenth and twentieth centuries was the *guaranteed* sight of the animal itself.[19] From the Society's perspective, seeing the animal was the very way in which the public would rather passively learn about the inhabitants of the natural world. Indeed, the act of observation was right at the heart of scientific endeavor from the Enlightenment onward. The dominant mode of display of animal life in bare cages over the course of the period decisively made the Zoo's creatures objects of total surveillance and so maximized the extent to which they could operate as informative representatives of species.[20]

As part of the effective communication of this kind of information, fragments of the once living were reconstituted and put on display. In many nineteenth-century zoos and museums, the display of skins, skeletons, and horns in close proximity to the living formed part of an exhibition designed to inform through the process of intensive observation. The Zoological Society of London viewed its menagerie and museum as closely related, animals moving in one direction between the two exhibitions in the dissemination of knowledge. Similarly, in the natural history displays at Bristol's civic museum, snails, frogs, and newts were displayed alongside the dead in the late nineteenth century.[21] Animal fragments were displayed at Bristol Zoo from very early in its history as key facets of the Society's educational performance. They were considered vital elements of the exhibition because they enhanced the observation of the captive wild by rendering animals absolutely motionless. In 1853, the skeleton of a lion, which had died of influenza, was rearticulated in the carnivora house, to be viewed together with living specimens.[22] Sometimes animal fragments also acted as substitutes for the living in the Zoo's comprehensive and representative zoological series. Stuffed chimpanzees, for example, were exhibited in the Zoo in the 1860s, allowing for the observation of the animal in spite of—or perhaps because of—the absence of a living example. Between the late 1850s and the mid-1920s the Society also operated a designated exhibition space for lifeless creatures above the Grand Refreshment Pavilion at the very center of the Gardens.[23]

Concurrently, the sight of the animal was necessary not only in order to effectively disseminate educational information but also in order to satiate visitor expectations.[24] After all, and as already noted, the generation of income through gate receipts was an important supporting pillar of the Society's zoological operation. The robust entanglement of these scientific and spectacular aims meant that the significance attached to *seeing* the animal persisted. The Zoo was no place for an animal to hide, and enclosure design over much of the period reflects this. The small and bare cages, dominant as a

FIGURE 13. The Polar Bear Court, c. 1930. (Roy Vaughan Collection, Bristol Zoological Society Archive, courtesy of the Bristol Zoological Society)

FIGURE 14. Ostrich enclosure, c. 1950. Harvey Barton & Sons Ltd. (Roy Vaughan Collection, Bristol Zoological Society Archive, courtesy of the Bristol Zoological Society)

FIGURE 15. Glass front big cat cage, c. 1950. Maine Photographers Ltd. (Roy Vaughan Collection, Bristol Zoological Society Archive, courtesy of the Bristol Zoological Society)

display medium in most zoos for so long, positioned animals at Bristol at the center of the visitor gaze, with nothing to detract from the panoptic observation of their form.[25] The bare exhibition space of the barred Polar Bear Court and the ostrich enclosure, for instance, were spaces that allowed for maximum visibility of the animals enclosed (figs. 13 and 14).

The twentieth-century introduction of barless enclosures such as the Monkey Temple and the rhinoceros paddock similarly privileged the sight of the animal. Both the lack of bars between the visitor and observed creature and the absence of obstacles in the enclosure ensured optimum visibility. The use of glass as a barless barrier in some twentieth-century displays further facilitated surveillance. Strong glass panels were placed in a small portion of the leopard cage in 1950 (fig. 15), after which a British Pathé newsreel reported "quite the latest thing in brighter zoos, [visitors] can now snap the man-eaters without bars to spoil that big game effect."[26]

Enhancing visibility was certainly one of a number of motives underpinning the creation of the innovative Nocturnal House in 1954. While the enclosure represented a move toward a zoogeographic display rationale, it was also pointedly motivated by the desire to bring animals of the dark into view through the artificial inversion of day and night: "There are many small and very attractive mammals which, being strictly nocturnal, are never seen by the public at zoological gardens. . . . The Council therefore asked its officials to try and experiment to ascertain whether by lighting their cages during the night and giving the equivalent of moonlight by day they could, as it were, be turned."[27] The house itself contained a number of enclosures, arranged in a semi-circular configuration, each lit by a light that mimicked the effects of sunlight reflecting off of the moon's surface. Double doors prevented unwanted flashes of daylight from disturbing the darkness and reducing visibility. The *Times* enthused that "it is safe to say that in no British Menagerie, at least, have kinkajous [small raccoon-like creatures] been seen to better advantage."[28]

The sight of the animal was not enough on its own to fulfill the Society's dual educational and entertainment agendas, however. As representatives of species, and as exotic attractions, creatures on display were meant to highlight the forms *and* behaviors of wild things. The perception of movement, so central a part of affective spectacle, and synonymous with animality in the celebrated early twentieth-century wildlife dioramas of Carl Akeley, for instance, was essential to the provision of the illusion of an encounter with *real* wildness.[29] Importantly, too, it was movement that decisively differentiated the experience from the observation of the no longer animate in museums of natural history.[30] The spectacle at the Zoo was fundamentally about watching animals *being wild, being animal.* In other words, it was about being convinced by a "staged authenticity."[31]

Thus, from its inception the Society energetically focused the visitor gaze not only on animal forms but also on their associated behaviors. In the process, they were able to illustrate the functionality of anatomical features while simultaneously heightening levels of amusement. The original Bear Pit, for instance, was constructed with the express purpose of exhibiting natural behaviors through demonstrating the readiness with which the animals ascend trees in their natural habitats. Similarly, the interior of the Monkey House at the end of the nineteenth century was "furnished with horizontal and swinging branches, chains and swings for promoting the incessant gambols of these . . . creatures."[32] Later enclosures often promoted animal activity through an aspiration toward the evermore-faithful reproduction of natural spaces, including in the replication of night-time conditions in the 1954 Nocturnal House. Indeed, one author wrote, "it is well known that nocturnal animals, generally speaking, are not very satisfactory exhibits in zoos as they are only to be seen running about and feeding when the visitors have gone home." The nocturnal house reframed nighttime creatures in captivity by giving visitors the chance to witness not only their forms but also their liveliness.[33]

Visitors responded to animal energy. There was, for instance, a distinct lack of visitor interest in the "listless and torpid" Mauge's opossum and wombat in the mid-1840s.[34] Their lack of animation lessened their capacity to amuse and inform. In 1879 schoolboy Frederick George Melville enthusiastically referred to the "quaint antics" of the monkeys, and Florence Grinfield remarked in 1894 that "the bear pit is visited by multitudes, and a rush is frequently made to see the bear on his Pole." A pool in the Polar Bear Court simulated the display of their aquatic behaviors, Grinfield noting that it was entertaining to watch the animal diving into its pool for a swim. Melville expressed his disappointment that he was not able to witness the bear swimming for himself during his 1879 visit. At the end of the 1930s, too, it remained clear that the agile acrobatic performances of the monkeys were "an unending attraction to visitors."[35]

The importance attached to the performativity of animal bodies—rooted in education and entertainment—stimulated the development of particular events that specifically displayed certain animal behaviors. Feeding time, for instance, was a moment at which creatures might be seen at their most *animal*. Watching animals eating has long been of intense interest to humans; a number of the earliest animal films presented in theaters were *actualitiés*, which showcased this behavior. People all over the world have enjoyed watching animals eating in zoos.[36]

Beginning at Bristol in 1845, keepers were ordered to feed their animals while they described the collection to attentive visitors. Feeding time, originally around the middle of the day (though this schedule was later subject to repeated alterations), subsequently became a principal visitor attraction. The manipulation of animals through the imposition of a regimented feeding schedule guaranteed the visibility and activity of some animals at certain times of day.[37] Grinfield poetically recounted the public significance of the lions' feeding time:

> And as the clock is striking the magic hour of *four,*
> "Feeding time," in the lion house, list! How the wild beasts roar.
> Off the Rink come the skaters,
> In the lion house they go,
> And "the young man from the country,"
> Who wants to see all "the Show,"
> With his pretty sweetheart too,
> And the children, of course, you know;
> And everybody in the grounds, they run with utmost speed,
> It is the feature of the day to see the lions feed.[38]

The feeding of these most charismatic of megafauna was not merely about nourishing visitors' ghoulish fascination. The act reflected perceptions of animals rooted in the convergence of education and amusement. The 1899 guidebook paid attention to anatomical adaptations and a hyperbolic spectacularization of the creatures' carnivorous proclivities: "It will then be readily observed how their entire structure is in keeping with their mode of existence . . . subsisting on the flesh of their fellow beings. Listen to the rasping, too, of the rough prickly tongue as it scrapes every particle of meat from off the bone. The vice-like grip of the front paws is also well illustrated . . . at intervals can be seen the soft elastic pads, on the sole of the foot, allowing their own to creep noiselessly upon the unsuspecting prey."[39]

Big cats aside, feeding time for the elephants, bears, sea lions, and great apes became central tenets of any visit to the Zoo across the nineteenth and twentieth centuries. Feeding schedules—and the event was staggered throughout the day for much of the period—were published at the front of the Zoo's guidebooks and on notices at the entrance lodges allowing visitors the opportunity of planning their day around such significant moments of animal action. The exhibition of the lively, frolicking, feasting beasts was achieved within the confines of designated animal spaces and so, in most circumstances, there remained a rigid spatial separation of entertaining and informative animal objects from visitors at the Zoo. This, in turn, reinforced the characterization of captive creatures as *objects to be looked at.* Nevertheless, this was just a single layer of conceptualization, a single facet of the human-animal relationships enshrined in zoo exhibition practices of the pre-1980 period. In addition, extreme proximity and multisensory interaction with the animated spectacle was of much significance, both to the Society and to the Zoo's visitors.[40]

Animals were often made to move beyond the structures—the bars, moats, or walls—that normally confined them in the creation of an occasionally intimate animal spectacular. The "thrill of proximity" was intrinsic to the appeal of animal events in the zooscape across the period.[41] Apes, for instance, often left the confines of their enclosures to perform in the Zoo's visitor areas. During the 1840s, the ungka-puti agile gibbon was removed from his enclosure and attached to a large tree by a long cord, allowing him to exhibit behaviors innate to his species. Removing animals, especially great apes, from their enclosures and parading them around the Gardens became particularly popular a century later from the 1930s. Vimy the wau-wau (gibbon) was often taken out into the Gardens while a young honey bear was walked on a metallic lead. During the 1970s and early 1980s, senior keeper Mike Colbourne often led chimpanzees and orangutans around the Gardens by the hand. While Colbourne viewed these brief excursions beyond the boundaries of their enclosures as a means of puncturing the monotony of the animals' daily lives in captivity, visitors were nonetheless thrilled by a sense of proximity to these perennially popular creatures.[42]

Beyond encounters like these, visitors often sought to get nearer for themselves by interfering in the managed habitats of the Zoo's captive creatures. The desire for interactivity was central to ways in which visitors sought to touch, feed, and otherwise harass the animals on display. Throughout much of the nineteenth and twentieth centuries, feeding animals at the Zoo was an integral part of an interactive and intimate encounter with the animal object. Such a spectacle was defined not by passive observation but by active and intimate, though intermittent, interspecies engagement. Prior to the 1930s, the visitor practice of feeding animals was implicitly condoned through official silence, partly in light of the enjoyment that visitors gleaned from the interaction. In 1889, it was enthusiastically reported that visitors fed coconuts to the polar bear through the bars of the Polar Bear Court. The animal then cracked them open on the stone floor before drinking the milk inside. Grinfield later noted that "tid-bits," like biscuits, fruits, buns, nuts, apples, and gingerbread, ought to be given to the animals. In fact, she strongly opined that visitor's bags be well stocked with such "goodies."[43] Stranger delicacies, too, were sometimes fed to captive beasts. In 1899 an ostrich was offered a sumptuous feast that included, among other things, penknives and iron coppers. It had long been erroneously believed that ostriches were capable of digesting iron. On dissection, a prayer book was discovered lurking among the bird's grisly innards.[44] After the 1930s, and commensurate with changing understandings of the workings of animal bodies (see chapter 3), measures were put in place that perpetuated the practice but that also stipulated which animals could and could not be fed.

Feeding was a moment during which visitors could initiate both animation and engagement in the animals they had come to the Zoo to encounter. This kind of relationship with animals could also show itself in moments of deliberate and sometimes profound violence, however. Such instances signify that some visitors related to animals as playthings not to be engaged with in the name of innocent spectacle (insofar as anything can be called innocent that puts something on display) but rather as objects to be exploited in the forging of an extreme form of display. The cruelty of young people, especially boys, toward animals was a frequent source of concern, especially during the nineteenth century. The Society's animals were spat at and teased, rubbish was regularly tossed into their cages, and catapults were fired at monkeys, seals, and waterfowl. Birds were mutilated, removed from their enclosures, or killed, and lit cigarettes applied to the noses of bears when they reached the top of their pole. This behavior was not uncommon in zoos and other animal attractions of the nineteenth and early twentieth centuries. The Society, as you might expect, considered this sort of conduct to be unacceptable, not least because it jeopardized the enduring vitality of its lively commodities, but also in the light of a strengthening public sensitivity to animal suffering, enshrined in legislation protecting animals in a variety of contexts, like the Cruelty to Animals Act 1876, the Wild Animals in Captivity Protection Act of 1900, and the Protection of Animals Act 1911, which consolidated aspects of previous legislation. Many of the perpetrators were banished from the Gardens.[45]

Reflecting developments at other zoos, cases of violent visitor behavior largely disappeared at Bristol after the Second World War, denoting a more widespread transformation in attitudes toward animal life among children, in particular, of the twentieth century (see chapter 4). Former head keeper Don Packham recalled, however, finding a monkey with blood pouring from its mouth sometime during the 1950s–60s. It had been fed an apple with a razor blade concealed within it.[46] For some, then, animals remained interactive and remarkably ambiguous targets for spectacular cruelty well into the latter half of the twentieth century.

The perception of animals as objects of interactive amusement reveals the existence of a multifaceted manufactured spectacle. Animal forms and behaviors were to be observed, encountered closely, and engaged with in a variety of ways, many of which were not directed by the Society itself. Yet, while these human-animal relationships remained predominantly rooted in the heightening of visual experience, animals at the Zoo were actually often presented and construed as far more than mere moving, silent bodies. The shrieks and howls of animals, and the enjoyment of activities that implicated touch, reveal that captive creatures were interpreted and presented as *multisensory* objects of information and amusement. Smells, meanwhile, were not infrequently eliminated from the zooscape. The multisensory spectacle was all about experiencing a supposedly authentic animal, but that animal reality could do without the associated stink.

The soundscape of the Zoo was an integral part of an informative and entertaining experience of wildlife, though of course it was far from static; the sounds of the Zoo would have been extraordinarily fluid, changing complexion as animals came and went, not only in and out of the Zoo, but also in and out of their living quarters. Nevertheless, some noises, like the roaring of lions, or the whooping of gibbons, were marked out as elements of the zoo's soundscape to be deliberately listened to.[47] The 1843 *Catalogue of the Animals* specifically directed visitor ears to the sound of South American howler monkeys: "The cry or mournful howl used by them in their native woods . . . is scarcely equaled by any known animal," it declared. Attention was also drawn to the melodious hoots of the ungka-puti: "the quality of these notes is very musical, though their loudness is somewhat deafening, but no description can convey an adequate or faithful idea of this musical effort" (i.e., you must go and listen to the creatures for yourselves). The importance of heeding the sound of the gibbons was noted once again in the 1957 guidebook.[48] Animal noises formed an important part of the visitors' sense of the zooscape. One visitor noted that a principal sound emanating from the Zoo was the bear "loading the wind with his tremendous howl," while Grinfield pointed in 1894 toward the sound of the roaring of the lions as darkness descended on the city.[49]

Alongside the visibility and audibility of the animals on display, their tactility was also significant. Touching is a way of knowing the world around us, and, indeed, it could be employed as a means through which the visitor could experience and learn about captive animals. Feeling the movement of animal bodies through the act of riding was one such tactile human-animal encounter. Riding on the backs of animals is, unavoidably, an instrumental use of their bodies.[50] The visitor enthusiasm for this practice is also a symptom, however, of the desire to *know* animal others through touch. Indeed, the Society was of the opinion that camel rides allowed children to experience "the sensation of camel locomotion."[51] The creaturely object of *encounter* was a moving, interactive being whose form could be felt as well as seen and heard. Just as at London Zoo, for instance, elephant rides had commenced at the Bristol Zoo by the later decades of the nineteenth century, but it was during the twentieth century, when the Zoo recast itself over time as an increasingly family-friendly space, that they proved to be most popular, particularly with children who were allowed to feel the movement of the massive creatures swaying beneath them. So important was riding to the provision of the animal spectacle that circus elephant Rosie was acquired in the late 1930s to join the younger Celeste, who had failed to "get the hang of giving rides."[52] Touch was also part of other animal attractions over the course of the period. Indeed, visitors simply sought to touch animals, despite the obvious risks involved, through the bars of their cages or over the other various kinds of barrier that were in place.[53]

Multisensory interaction, then, was fundamental to the complexion of human relationships with amusing, interesting, and supposedly authentic animal exhibits at the Zoo during the pre-1980 period. Significantly, however, very rarely was attention explicitly directed toward animal smells, and, indeed, visitors often deemed such olfactory perceptions offensive. Animals as objects of amusement were not meant to be unpleasant, even if the censorship of smell negatively affected the supposed authenticity of the creaturely encounter. None of the guidebooks explicitly directed attention toward the aroma of the animals, and, indeed, the innovation of glass enclosures in twentieth-century zoos not only facilitated a clearer, more proximal view of the animals exhibited but also served the purpose of lessening the invasive impact of odors.[54] Big cat enclosures had to be regularly scrubbed clean in order to eliminate the stench of ammonia from urine, and the Bear Pit was renowned among visitors for its foul smell.[55] More specifically, a porcupine was removed from the Nocturnal House in 1955 to eliminate the odor it produced.[56]

Ethologist Heini Hediger noted that bad odors in captive spaces were increasingly considered evocative of unacceptable conditions. Certainly, some visitors appear to have equated the cleanliness of enclosures at the Zoo with high standards of animal living. It is no surprise, then, that those expressions of distaste in relation to odors increased during the twentieth century, when important sociocultural changes came to have substantial impacts on the ways in which animals were construed and presented as objects of entertainment.[57] It is to the impact of these changes on the instrumentalization of animals as performers on the zoo stage that we now turn.

TOWARD A HAPPY, GREEN ZOOSCAPE?

These ways of presenting and relating to captive beasts remained, at their heart, remarkably static throughout the nineteenth century and for the better part of the twentieth century. Linnaean taxonomy, increasingly coupled with a more zoogeographic arrangement, was fused with the enduring need to render the animals visible, animate, and accessible in a multisensory way. Various conceptualizations of animals converged to create a multilayered exhibition rationale that resonated with the desires of paying visitors and that was also conducive to the Society's principled educational imperatives. Over the course of the period, however, and particularly post-1900, the Society was forced to respond to changes in cultural attitudes toward the treatment of captive creatures. These shifting attitudes toward nature contributed to transforming expectations of animal encounters and this, in turn, inspired a series of reinventions that eventually culminated in a sweeping revolution in the way in which animals and animality was staged in the final decades of the twentieth century.[58]

These changes have their roots in the nineteenth century. By the end of that century, a rising critique of animal captivity was an important symptom of an intensifying sentimentalization of animals and sensitivity toward their exploitation in Britain (the full impact of this is explored in chapter 4). The Bristol and Clifton Anti-Vivisection Society was established in 1883, for instance, and was one of many such organizations throughout the country. Each of these sought to put an end to the "cruelty and cowardice" of the practice on "dumb beasts."[59] From the turn of the twentieth century this rising concern for animal exploitation began impacting on display practice at the Zoo. In 1906, the Society received a letter that condemned the overcrowded conditions in which animals were maintained. Later that year, it held a fete in the Gardens specifically for raising funds for the improvement of animal accommodations.[60] Indeed, animal attractions elsewhere had already responded to critiques of captivity. In 1896 German animal and ethnographic showman Carl Hagenbeck patented the naturalistic animal "panorama" at his Tierpark in Stellingen near Hamburg, in Germany. This method of display presented animals as ostensibly at liberty. They were no longer *explicitly* confined by the stark structures that had characterized animal displays in nineteenth-century zoos and that increasingly signified imprisonment and unwarranted punishment. Instead, creatures on display were separated from visitors only by a moat. In so doing, the display structure rendered the animal spectacle itself palatable and, of course, more extravagantly performative because it positioned animals in an artificial environment that appeared to be approaching those in the wild places themselves. The shrinking of the distance between the captive and wild states assuaged concerns, to some degree, about animal suffering. The "Hagenbeck revolution" was, at its core, about the depiction of harmony and happiness in a world in which animals frolicked and that people could only briefly glimpse.[61]

In the light of financial pressures, it took a number of decades for these sentiments to impact on the zooscape itself. Bristol Zoo's first enclosures that did away with explicit signs of imprisonment were the 1928 Monkey Temple, the 1930 raccoon pit, and the 1931 outdoor reptilium. Each of these enclosures presented animals without the intrusion of bars in spectators' line of sight. In so doing, they produced an illusion of freedom.[62] The rationale underpinning the 1935 polar bear ice floe enclosure also rejected traditional Victorian structures of confinement. At the same time as it contextualized the species within a frozen world it partially obscured the structures of captivity by fabricating an artifice of freedom that was more conducive to the visitor tastes of the day (fig. 16).

This theatricality, and the associated illusion of freedom crafted through the removal of overt signals of captivity, created a spectacle that sat comfortably with shifting public sensibilities in the early decades of the twentieth century. As indicated in the previous chapter, however, attitudes to wildlife and wild

FIGURE 16. Polar bear ice floe exhibit, 1967. S. W. Bowden. (Roy Vaughan Collection, Bristol Zoological Society Archive, courtesy of the Bristol Zoological Society)

places underwent a series of shifts over the course of the twentieth century, and particularly in the liberalizing world of the postwar decades. By the later years of the twentieth century, such attitudes had gathered strength, and opposition to animal captivity had significantly intensified. One correspondent to the Zoo complained in 1968 about "the ghastly Victoriana still in being at Clifton." The cat houses, Bear Pit cages, and "even the fairly modern monkey 'prisons' are all relics of a bygone age," it continued. These kinds of sentiments gathered force during the years that followed.[63] Commercially speaking, at least, it was vital that the Society respond to reconfigurations in attitudes toward nature by remodeling its theater of animality along more agreeable lines.

Thus, many of these Victorian relics were removed and replaced and the zooscape gradually remade. The Bear Pit, for instance, had remained almost entirely unchanged for over 145 years but was converted into a series of enclosures housing smaller mammals in 1971. The spectacle of a bear ascending from the pit had become too reminiscent of archaic animal pastimes now predominantly associated with spiteful and sadistic propensities rather than "innocent" spectacle. Owing to various constraints, including the cost and feasibility of building improvements, wholesale change took time, and thus the zooscape's mosaic of enclosure styles shifted in complexion like a Rubik's cube rather than wholly transforming overnight. The tiny enclosures for big cats were enlarged only in 1977, and only in 1983 was the primate house reconfigured into something less punitive in appearance.[64]

Even these enlightened reconfigurations were not to stand the test of time. By the end of the twentieth century, an ever-intensifying concern for the state of the planet and the lives of its wild things inspired a further revolution in enclosure style. The naturalistic and barless Monkey Temple, with its concrete pit that had been so innovative for its time, and that had been built in some accordance with the progressive barless principles that underpinned the Hagenbeck revolution, had become "an embarrassment." Its concrete pit was sterile and small; the so-called naturalism of the late 1920s was much removed from the new naturalism of the late twentieth century.[65] The latest innovation in zoo display—zoogeographic immersive environments—thus replaced some of the older enclosures. Exhibits such as Seal and Penguin Coasts and Zona Brazil formed part of a more palatable animal exhibit by obscuring the structures of captivity even more than the barless enclosures of an earlier age and by further shrinking the distance between "here" and the exotic "there" through the presentation of naturalistic environments that appeared to replicate wild worlds more convincingly than ever before.

This obfuscation of the structures of captivity and the reconfigured attitudes toward nature that it represented was coupled with a transformed educational mission rooted in the ethos of conservation. The postwar rise of television as a mode of entertainment and education in the home, and of natural history film, which created an illusion of a world apparently *as it was,* from the BBC's Natural History Unit in Bristol, for instance, further diminished the role of zoos in presenting the forms and behaviors of animals to the public. Instead, zoos became places where people could learn about their own behaviors and the roles they played, alongside animals, as constituent parts of an increasingly fragile and entwined wild world. The vicarious jaunt into a wild and distant nature via the observation of instrumentalized animals in zoos had become, in theory, one drenched in guilt and personal responsibility. Immersive exhibits not only positioned animals as performers on a more elaborate stage than ever before, but they also deployed the animals and their worlds in order to gesture toward the significance of ecological awareness, the importance of animal welfare, and the urgency of conserving wild worlds all at once.

Indeed, in immersive exhibits, animals were contextualized in habitats that were meant to be more authentically wild than ever before. Twilight World, for instance, displayed a range of nocturnal creatures in reduced light and in a variety of common environments, including a town house and a desert. It was a substantial reimagining of the 1954 nocturnal house. Seal and Penguin Coasts likewise displayed these creatures in a cold coastal habitat, and Zona Brazil exhibited together a range of South American animals in a walk-through rainforest setting. These exhibits imposed order on the animal world, just as the very first nineteenth-century enclosures had done, but they did so in order to portray, not a catalog of taxonomies, but a rich biosphere, with its inhabitants living in relationship with one another.[66]

In so doing, the Society mediated the meaning of the animals in its care to communicate a pedagogical message rooted in conservation and ecology. As visitors travel through the wild places reproduced at the Zoo, they encounter a spectacle reflective of a changed conceptualization of the human place in the world and its relationship to animal life. Immersive exhibits allowed visitors to enter the exhibition space, enhancing the illusion of authenticity and widening the performance arena in the apparent convergence of human and animal spaces.[67] In contrast to earlier modes of display, the visitor is no longer positioned as an outside observer of an isolated animal world but is, instead, integrated as part of a larger, more evidently connected, biosphere.

The precise form of the human-animal relationships enshrined in these kinds of exhibits is sharpest in the darkness of Twilight World. With a wide range of taxa tightly integrated into an exhibit rooted in an ecological conceptualization of the natural world, the theatrical display features four specific habitats united by their animation once night falls. The visitor's journey of discovery begins in the desert where, from the sanctum of a field station and in the silence of dusk, "the desert comes to life in front of you" as mongoose rush about their sandy domain.

Moving onward, the visitor encounters a forested world where sloths, jumping rats and bats, and assorted Australian fauna can be viewed from the seclusion of a blind and under a starlit sky (fig. 17): "we are truly in the forest and 'feel' the animals around us." While the animals here have been separated according to species, this separation is integrated into a display that creates the illusion of free movement within a single sylvan habitat. The walls of the display area are smoothed in order to remove brickwork and to evoke natural surfaces so that the sense of touch cannot wholly betray the artificiality of the space.

FIGURE 17. Forest habitat in Twilight World, c. 2000. (Bristol Zoological Society Archive, courtesy of Andre Pattenden)

Leaving the forest behind, the visitor descends beneath the earth into the dark domains of blind cavefish and naked mole rats. As the end of the journey, the visitor emerges into the more familiar environs of a town house inhabited by creatures, like spiders, rats, pet fish, and cockroaches, with which people, often reluctantly, share their homes. From beginning to end, from light into darkness and re-emergence into the town house at dawn, the visitor is transported into the shrouded places in which some animals live their lives and make their homes. This form of instrumentalization of animals through spectacle is rooted in the creation of a living, breathing ecosystem where the glass barriers separating human visitors from captive animals are obscured and in which visitors are supposed to discover the creatures of the darkness for themselves by wandering through what appear to be wild places. Visitors are positioned as travelers in a secluded animal domain usually existing well beyond the gaze of sleeping people. Yet, at the same time, there is a deep sense of shared space. This is a world whose soils bear the footprints of both humans and animals. Information about the animals is presented through human artefacts, such as notebooks, maps, and field guides, while the town house habitat, in particular, depicts a communal environment in which our homes are quite decidedly not only ours at all.[68] This ecological presentation of animal exhibits represents the latest incarnation of re-created nature at the

Zoo, where species barriers are broken down and where humans and vulnerable animals share the spaces and places of the Earth.

Despite these shifts, however, there remained remarkable continuities in the ways in which animals were deployed as objects of education and amusement through display, even in the age of zoo-based conservation and transforming expectations of captive institutions. The Zoo remained a tourist attraction that was obliged to deliver a product—living, breathing, wild things—for the pleasure of its visitors. More than this, however, its conservation-based educational mission absolutely demanded unmediated encounters with *real* animals in order to differentiate the experience from one arbitrated by the now ubiquitous television or, later, the computer/smartphone screen. Thus, the guaranteed sight of animated animals remained critical to the Society's and its visitors' relationship with its animal performers. In the midst of extensive landscaping of immersive exhibits, which gave animals ample opportunity to evade direct observation, the Society continued to develop techniques that guaranteed visitors an extended glimpse of the captive wild; the island enclosure of mongoose lemurs, for instance, was significantly redeveloped in 2000, in substantial part to improve visibility. Use of glass barriers was another strategy employed with increasing regularity over the course of the past few decades.[69] The rise of Internet technology, too, opened up new and exciting ways of watching wildlife. It maximized the visibility of captive animals through intensive surveillance. Closed-circuit cameras positioned in the gorilla enclosure, in Seal and Penguin Coasts, and in Meerkat Lookout (2011) broadcast live feeds of the animals via the World Wide Web so that the public might have the opportunity to observe them in their heavily landscaped captive worlds from the comfort of their own homes, day and night.[70] This is intrusive spectatorship and vicarious travel at the extreme and testifies to a perpetuation of a desire to objectify animals and position them in ways that could provoke curiosity and potentially legitimize voyeuristic consumption.

Continuity in human-animal relationships also lurked beneath surface-level transformations in the character of multisensory encounters. Sounds and smells remained of interest and revulsion, respectively. A desire for contact through feeding also endured at the heart of the zoo experience. Over the course of the second half of the twentieth century, there was greater recognition of the damage that could be done to precious beasts through the practice of public feeding. From around the middle of the twentieth century public feeding practices had come under increasing scrutiny in many of the world's zoos. In 1951, for instance, the Society lost its entire stock of gibbons. They had succumbed to a human form of dysentery, which had been transmitted through their ingestion of contaminated food.[71] A tension between the conceptualization of animals as valuable commodities *and* instructive and thrilling spectacles had emerged because, as noted, large animals were becoming harder to obtain through the channels of the wild animal trade. This combined with rising concerns for animal welfare in British culture at large, alongside a deepening understanding of the intimacies and fragilities of animal bodies (more on this in chapter 3), to affect exhibition and interaction practices that had endured over the vast majority of the Zoo's history. In 1967, honorary pathologist T. F. Hewer reported that "at present, members of the public feed our monkeys, bears and many other animals without any control. On my own observation, a high proportion of the visitors, especially the children, feed them continuously." Consequently, all public feeding was prohibited from 1 January 1968. This ban met with a mixed response, reflecting an enduring visitor desire to interact with animals as part of the theatricality of their visit. The possibility of feeding animals, however, remains among the anticipated animal encounters during a day out at the Zoo today. The need to use signage to prohibit this behavior testifies to the endurance of that particular practice.[72]

While visitor feeding as an essential facet of multisensory spectacle was prohibited by the late 1960s, the exhibition of feeding time as an organized event in the daily schedule nonetheless remained intact. The display of such fundamentally *animal* behaviors, however, underwent important changes that reflected an increasing concern for the replication of the natural rhythms of the wild. Since the end of the 1970s, the public exhibition of the lions' feeding time took place at an unspecified hour, rather than at a predetermined time, in order to better replicate the irregular feeding habits of these animals in the wild. That being said, however, the feeding of many other animals in the Zoo, from gorillas to seals and penguins, for instance, today remains tightly integrated into the daily schedule of events because they endure as moments of guaranteed animality.[73]

The popular pastime of riding the animals also became less tolerable in the second half of the twentieth century. After the demise of Rosie in 1961, no

elephant was again ridden in the Zoo Gardens, and camel rides and llama-
and goat-pulled carriage rides had disappeared by 1964. Although the rationale underpinning the end of this practice is uncertain, it reflects a wider
trend, since the Zoological Society of London also abandoned elephant rides
in 1960.[74] In these ways, the animal spectacular was reconfigured, not only in
the interests of preserving and enriching increasingly rare and thus increasingly ecologically valuable creatures but also as a way of protecting people.
This exemplifies a significant tension between two distinct but connected
layers of relationship at the Zoo: conceptualizations of animals as objects of
science and spectacle.

Yet, the importance of touch, of relationships with animals mediated by
corporeal sensation formerly embodied in feeding and riding practices, survived these changes. It exists as a controlled but nonetheless "unforgettable
experience" in the Feed the Lorikeets exhibit, where visitors can feed nectar to
the colorful birds by hand. Furthermore, as a way of coming to know animal
life, touch has also come to form an increasingly pivotal mode of pedagogical
instruction in the age of zoo-based conservation, even though changing enclosure design somewhat limited the availability of casual tactile encounters
of earlier years. Education officer Simon Garrett noted in 1993 that in close
encounters with a rat, chinchilla, frog, hissing cockroach, or snake, touch
was a vital component of the children's educational experiences. It is in this
context that death, too, retains its place in exhibition practice. Silent remains
of the formerly living transmit a range of key biological, zoological, and conservation issues such as evolutionary development and animal adaptations.
A giant octopus preserved in formalin, the mounted head of a flamingo, the
skins of a zebra, gorilla, white tiger, and okapi, exoskeletons of spiders and
scorpions, and skulls of a tapir, peccary, and chimpanzee all sit in the class-
rooms of the Education Building.[75] Touch, in tandem with the enduring supremacy of vision, allows children to forge connections and to access what is
presently considered the realities of animal life.

CONCLUSION

Bristol's zooscape is, and has always been, a mosaic. It is, in both physical and
metaphorical senses, a patchwork of ideas about and attitudes toward the rest
of the natural world and the beings who dwell within it. As a public space,
performance and theatricality have forever been at the heart of what it has

been about. And performance and spectatorship have also been at the core of our interactions with animals in modernity. Over the course of nearly two hundred years animals were presented to the people of Bristol and beyond as multisensory objects of education and entertainment. The instrumentalization of animals as display objects was itself, however, complex. While the guaranteed sight of animal life in captivity was critical, wildlife also needed to be vibrant and open to interpretation through sound and touch and, to a far lesser extent, smell in order to effectively serve in the communication of educational messages and the provision of (usually) innocent pleasures. As part of this, animals were transformed into ciphers. They were positioned in an array of contexts in order to satiate the need to entertain and to improve, and this manipulation and instrumentalization of animal life has remained remarkably consistent over the entire lifespan of Bristol Zoo.

Over the years, the blend of contexts within which animals were displayed transformed not only in the light of changing public conceptualizations of animals and environments but also as a result of changing technologies and increasing awareness of the place of humans in a vulnerable biosphere. After the turn of the twentieth century, the bars came down and glass or open air replaced them. Linnaean taxonomy was eventually subsumed by zoogeographic displays, and these in turn ultimately made way for immersive exhibits where the boundaries between human and beast were to some extent dissolved. Animals and their represented worlds served as the figurative message in a bottle through which words promoting responsible attitudes and actions toward nature could be delivered. Entertainment and education—and the objectification of animals within those contexts—took place across a zooscape whose tectonics were perpetually shifting.

However, the creation of spectacular stages was, of course, also about finding ways of keeping precious beasts alive. Animal houses—rising and falling—were a single part of a great experiment that lasted across nearly two centuries. These creatures, and their habitats, were variously valuable objects, not only of instruction and amusement, but also of experimentation too, for in their flesh and bones they contained a world of secrets.

3

The Great Experiment

Across nearly two centuries, zoo animals were understood and objectified in varying ways as precious commodities and entities of amusement and instruction. However, their bodies—inside and out—were also framed according to the secrets they held: secrets to the unlocking of knowledge about the natural world and secrets about how to keep precious creatures alive in captivity. Studying animals, watching them, and delving deep into their bodies was an essential layer of human-animal relationship at the Bristol Zoo, and this transformed them into objects at the heart of what was dubbed a "great experiment." This experiment has lasted across nearly two centuries, and the knowledge gleaned from these interactions has had substantial implications for the management of animal life, in captivity and beyond. In the 1950s, for instance, keeper Stan Puddy could be found working in the primate enclosures, his face covered by a mask to prevent him from transmitting influenza to the animals he cared for. The cage design itself guarded against trans-species transmission of other respiratory diseases such as tuberculosis, too, by separating the display space from the observation areas from which visitors peered in.[1] Enclosure innovation and husbandry practices such as these were founded on the realization of a deep physiological resemblance between human and nonhuman primate species, which had itself come about as a result of a long scientific tradition of observing animal bodies. Earlier practices such as the provision of open spaces through the fresh air and sunshine treatment for tuberculosis, which had been used in the medical profession from the end of the nineteenth century and then applied to animals, was also focused on that recognition of resemblance and, in most circumstances, sat alongside spatial separation as established approaches to animal bodies at the Zoo in the mid-twentieth century.[2]

As captive animal bodies yielded their secrets through observation on the ground, zoological understanding grew in both volume and depth. Medical technologies developed over the course of the twentieth century (especially), and this led to the adoption of new husbandry techniques. Between 1972 and 1980, Honorary Pediatrician Dr. Beryl Corner, a pioneer of neonatology in Britain, assisted in the provision of neonatal care for young primates housed at the Zoo. On one occasion, she used an incubator that was normally reserved for human children at Southmead Hospital in North Bristol to provide care for a new-born chimpanzee.[3] Over time, increasingly complex human medical procedures were adopted. A cataract was removed from the gorilla Romina's right eye in a revolutionary procedure performed in 2002 by ophthalmic surgeons from the Bristol Eye Hospital, whose jobs were normally to conduct such operations on human patients. Four years later, female gorilla Salome gave birth to a youngster following a prolonged period of treatment for infertility utilizing drugs normally prescribed for the treatment of women, while gorillas were born in 2011 and 2016 by emergency caesarean section performed by surgeons from Bristol hospitals.[4] Thus, over the course of half a century, animal husbandry transformed not only in light of an increasingly sophisticated armory of medical techniques but also an ever-heightening medical and zoological knowledge gleaned, in part, through the uncovering of the secrets of animal physiologies. These processes were founded on the extraction of information from animal bodies.

This was an important layer of the Society's conceptualization of animals from its inception. At the first meeting of the Society held at the Bristol Institution—that center of scientific endeavor at the heart of the city—on 22 June 1835, it was indicated that the Society would look to "promote the diffusion of useful knowledge by facilitating the study of the habits, forms, and structure of the Animal Kingdom."[5] Throughout the near two centuries since, the Society was consistently explicit about this intention, even while the character of the knowledge it sought to extract and then deploy shifted in significant ways. According to this educational and scientific paradigm, animals were subject to an intense medical and zoological gaze. Captive creatures held mysteries in their flesh and bones and the extraction of the solutions to these enigmas was part of a magnificent experiment spanning across modernity and which endures to this day.

This practice of experimentation, and the governance of animal (and human) bodies that invariably follows, is a significant layer of human-animal relationship that sits right alongside interactions rooted in the valuation and display of animal bodies in British culture. We need only consider the ways in which we have tested cosmetics and pharmaceuticals on animals to understand both the significance of our desire to interrogate lively beings and the variety of methods we devise to make use of that knowledge. Indeed, in 1954, the Society's mission was emblazoned on its newly created coat of arms. At its heart was a maxim that epitomized this way of relating to animals: "Ask of the beasts and they shall teach thee."[6] Asking searching questions of captive animals was an essential practice, for only through the observation and examination of animal bodies and behaviors could the Society learn how to effectively maintain its valuable animal commodities in captivity. As we have seen, the dual imperatives of science and commerce combined to render the vitality of the beasts a priority—actually an absolute necessity—from the outset. The accumulation of knowledge through perpetual experimentation had the effect of transforming animal management, not only in terms of the provision of appropriate enclosure space but also in the coordination of feeding, breeding, and, in the end, the very preservation of vitality itself. Such knowledge was not a secret ferociously guarded by the Society, of course. The dictates of nineteenth-century "rational recreation" and the collaborative scientific educational paradigms that followed over the course of subsequent decades compelled the Society to circulate its findings widely. This included dissemination among the zoological community, and this meant that the "great experiment" affected animals well beyond the Zoo's own physical boundaries.

These kinds of interaction with animals reflected not only a culture of objectification—and thus of human separation from the rest of nature—but also the homogenization of individual creatures into species groups. In the process, this way of knowing implicitly obscured the presence of individual internal animal viewpoints. Consequently, the objects of this zoological gaze were cast within an almost-Cartesian, mechanomorphic framework, as animate machines. As objects of experimentation, the animals construed in this layer of relationship were viewed as *predictable* products and model exemplars of natural phenomena and processes.[7] Thus, individual captive animals were positioned as embodied gateways to knowledge because to understand the individual was also to understand the fundamental nature of species. At the same time, of course, to understand the species was also to understand the individual, and this permitted more effective management of animals in captivity. If we look at guidebooks, which were often at least partially written by men working in the scientific professions and which were primarily informative in tone, we can clearly identify this interaction between knowledge of the species and knowledge of the individual. They were mutually constitutive.

The necessary focus of this gaze meant that animal bodies were rigidly, and increasingly, controlled in order to expedite the extraction of knowledge. There were a number of ways in which this manifested. Over the entire period, the Society and, later, animal keepers maintained extensive zoological records produced through the rigorous observation of captive animals. Over time, these records became more thorough, and this helped facilitate an increasing understanding of bodies and behaviors. The gathering of this knowledge, in turn, facilitated the development of new kinds of zoological practice. Over the course of the period, a progressively more refined armory of interventions could be deployed. Animals were intensively studied as they breathed and birthed in their captive spaces.

Excretions were examined; fluids extracted and analyzed; and bodies themselves invaded in order to uncover the keys to an ever more extensive zoological comprehension. Indeed, the accumulation of knowledge gradually permitted the development of preventative as well as reactive medicine as a way of protecting animal health. When treatments and preventative measures fell short, however, intensive surveillance of dead animals made them important zoological resources in their own right. Evolutions in scientific practice meant that postmortems became both more frequent and also better able to disclose not only the cause of death but also wider truths about animal bodies, and this information fed into a growing body of zoological knowledge. The accumulation of this knowledge had ramifications for animals in all sorts of environments, not only those residing within the artificial spaces of the Zoo, but also those in the captive and wild worlds far beyond.

INFORMAL EXPLORATIONS

To begin with, and across the nineteenth and early twentieth centuries, zoo-logical knowledge was in large part only skin deep and its accumulation only quasi-systematic. Guidebooks are a useful repository of information regard-ing the state of this knowledge. Though written for the visitor rather than the high-minded natural scientist, they give a sense of what was considered important about animal physiologies at any given point in time. Guidebooks from the 1800s reflect a slightly shallow conceptualization of bodies and be-haviors, relying on anthropocentric anecdotal contexts coupled with a con-centration on the physical relationships between the apparently diametrically opposed human and animal worlds. This was a result not only of an incom-plete (by our standards) comprehension of the workings of animal bodies, but also of the preeminence of a scientific gaze rooted in Linnaean taxonomy and the enduring sense, even post-Darwin, of an essential human superiority over the multitudinous occupants of the rest of the natural world.

The 1843 *Catalogue of the Animals* emphasized a deep-seated antagonism between monkeys and the humans they lived alongside, particularly in the context of the Indian subcontinent. Monkeys were also rudimentarily distin-guished from great apes and baboons, and from each other, according to their morphologies and, in particular, the structure of their tails. Certainly, these descriptions were intended to enable both the learned and the ill-informed observer to gaze upon the forms of these animals, to absorb information about them, and to then place them within a zoological series—a chain or ladder of being—that highlighted both continuity with and differentiation from creatures similar to them in form. More than that, however, such de-scriptions locate animal identity in their predictable bodies and the physical relationships between species.[8]

This chain or hierarchy of being implicitly included God in what was a Creationist scientific paradigm. In the earliest years, before Darwin (and others) circuitously questioned the role of a creator in the forging of species, chimpanzees were described according to their inability, unlike the excep-tional human, to turn their heads toward the heavens. By the 1880s, God and the heavens had disappeared from the guidebooks' scientific discourse. Yet, they continued to refer, for instance, to the "monstrous" serpents of the Amazon and their place in human superstition and ritual.[9] This reflects a perpetuation of an anthropocentric scientific model through which animals were understood.

On the ground, zoo staff maintained records about the Zoo's captives, and they show that, like the guidebooks, only surface-level workings of individual animal bodies were known or considered important to those—the Society—who were directly concerned for their welfare across the nineteenth and early twentieth centuries. From the outset, rudimentary details were recorded in committee reports, reflecting the informal character of animal maintenance, resulting from a surface-level comprehension of animal bodies. Subsequently, for the first century or so, keepers operated in a broad capacity as glorified housekeepers, carrying out the Society's bidding. They tended not to have received formal scientific training, and many had no experience of working with animals at all. The first "beast keeper," Harry Richardson, had been with Stephen Polito's popular menagerie on The Strand in London until 1814 and had operated his own exotic collection before he started work at Bristol Zoo in 1836. The General Committee granted Richardson much responsibility; he was able to acquire animals, oversee their care, and manage staff below him. But he was not charged with maintaining scientific records about the animals he managed: that was the Society's prerogative.[10] The Society itself was pre-occupied with recording practical details such as physical relocations, births, and fatalities rather than the intimate workings of complex forms.

Medical practices similarly reflected this comparatively rudimentary comprehension of animal bodies. Veterinary medicine, which is, of course, reliant on, and the product of, a comprehension of the workings of animal bodies, was actually relatively well established by the mid-nineteenth century. The first modern veterinary school had opened in Lyon in 1761, and the Royal Veterinary College, London, had been founded in 1791. Early veterinarians were principally concerned with livestock; they were "physicians of the farm."[11] Exotic beasts were another matter entirely. Despite increasing professionalization, veterinary medicine during the first century or so of the Zoo's operation remained rather informal. A mixture of professional and nonprofessional men (quacks) practiced it, for there was no effective regulation of the profession at the time: anybody could practice on an animal body if they so wished.[12] Indeed, given the lack of institutional knowledge of exotic animal bodies, the dearth of available pharmacopeia, and the absence of regulation of a body of professionals charged with acquiring and then using what knowledge did exist, it is unsurprising that zoos initially tended to rely so heavily on local knowledge acquired from people who usually lived in close proximity to exotic animals. Such people possessed specialist knowledge gained through habitual exposure and, as described in the introduction, often accompanied large animals on their journey to the Zoo in order to educate inexperienced zoo personnel.

The first person specifically employed to treat sick animals at Bristol Zoo was a Mr. Webb. Webb was appointed some fifteen years after the employment of the first zoo vet (actually termed "medical attendant") at London Zoo: the influential veterinary surgeon Charles Spooner.[13] Accompanied by his family, for he was required to live on-site (like many keepers), Webb had immediate access to the animals of Bristol Zoo twenty-four hours a day. For a short time around 1844, he "gratuitously" administered medicines to some of the Zoo's animals in response to the observations he had made about their bodies.[14] Standard veterinary interventions of the time, which were stenciled directly from the blueprints of human medical practices, included treatment with antimonials (used as emetics to purge the body of toxins through vomiting), mercury (used for a variety of reasons including as an antiseptic), digitalis (used in the treatment of heart conditions), and bleeding. These were typical interventions in the "age of heroic medicine" during which purging the body of impurities was often the principal aim.[15] Webb's was an informal mediation, and it was shortly followed by the creation of a designated medical space. Emulating practice at the Zoological Society of London, in 1864 a space that could be used to isolate and treat sick animals was created.[16] Bit by bit understandings of the needs of animal bodies and the provisions necessary to diagnose and prevent ill health developed. What was referred to as a Watson's Ventilator, a mechanical instrument that circulated the air in enclosed spaces to remove foul vapors—miasmas—from, for instance, mines, sailing ships, and public buildings, and an operating den were added to the Zoo's medical facilities in 1870 and 1874, respectively. They were part of a growing behind-the-scenes "invisible zoo" dedicated not to theatricality but to management.[17] Thus, as technologies and understandings of animals in captivity evolved over time, layer upon layer were added to the strata of interventions available in the provision of animal well-being.

Carefully managing animals as scientific specimens was not merely a matter of protecting valuable creatures and the associated accumulation of knowledge, however. The ability to banish death was a measure of the Society's expertise. As already described, in early zoos keeping animals alive in artificial environments was a substantial and often unsuccessful undertaking. In light of this the Society measured its own success against conditions in other captive spaces, and this meant that there was more at stake in the extraction of animal secrets than merely effectively managing captive beasts. In 1851, for instance, high levels of mortality were very much considered to be acceptable when compared to similar rates in other menageries. A decade later, in 1861, it was reported with palpable satisfaction that the entire animal collection had passed through a severe winter unscathed (by natural causes), whereas zoos in London and Hull and all the traveling menageries had purportedly suffered unprecedented fatalities.[18]

Despite all of this, however, "death [was] busy in the Zoo." There was a profound likelihood that animals, which had been featured in official and unofficial guidebooks, might have long expired by the time visitors got around to making the trip to the living exhibition at Clifton.[19] Until the turn of the twentieth century and the rise of antibiotics, the medical interventions described above were impotent against bacterial infection, the major cause of disease and death among humans as well as their animal wards.[20] Inevitably, and despite the best efforts of the Society and its staff, the nineteenth- and twentieth-century zoo was a place of perpetual death through diseases that we now consider eminently curable. Yet, even after death many animal bodies served important purposes in an expansive experiment. Their last mortal breath did not signal the expiration of their contribution to the accumulation of scientific knowledge. "Dead animals . . . tell tales," argues Stephen T. Asma. Theirs is a "mute testimony. . . . Only in death [do they] pause long enough for our analytical minds to torture some truths out of them."[21] Thus, responses to death, through postmortem investigation or the formation of investigative committees, occupied a prime position in the machinery of animal husbandry. Such procedures allowed for the identification of disease in the dead, and the secrets that they unveiled increasingly permitted the effective management of still-living animal bodies. Postmortems were taking place at the Zoo from the earliest days. In 1847, for instance, a "medical gentleman" attributed the cause of death of a female tiger and a female leopard to "natural causes."[22] From that year onward, and reflecting the practice of civil registration of human deaths introduced in July 1837, details surrounding the circumstances of each animal fatality were officially recorded and reported at every monthly meeting of the General Committee so that patterns of mortality could be tracked across the collection and over time. Moreover, when mortality was unusually high, special subcommittees were convened to ascertain the causes and suggest potential solutions. One such subcommittee was formed in response to an unspecified spate of deaths in 1861, for instance.[23]

Dead bodies were deployed in intensive study not just to understand the cause of death but also to uncover a vast world of bodily secrets. In this context forms, bones, blood, and soft tissues were transformed into objects that held the keys to human understanding of the workings of vibrant animal bodies. Indeed, they were explicitly considered as reserves of biological information, waiting to be used."[24] The extraction of this kind of knowledge was considered important from the outset, not just for the Society's own captive stock, but also for the advancement of zoological knowledge more broadly. In 1841, a female chimpanzee died at the Zoo. After presentation to the Bristol Institution, one of its members, Dr. Alexander Fairbrother, a member of the Royal College of Surgeons who had been appointed to the staff of the Bristol General Hospital in 1838, dissected the cadaver. The scientific examination of this animal's body was employed as a means of investigating the anatomical and ontological similarities and differences between human and nonhuman primates, which so preoccupied nineteenth-century minds. When examined, the body was found to be remarkably similar to that of a human. Nevertheless—and of course prior to the emergence and dissemination of Darwin's insights—it was evident that it lacked the apparatus of speech; this, in combination with the assured absence of an immortal soul, confirmed that there was a fundamental distinction between humans and what we now know to be other species of primate.[25]

The increasing ability of the Society and its keepers to delve into and thus understand the forms and behaviors of animals, even if only in what we now consider a wholly rudimentary fashion, had important consequences for the lives of captive animals. To realize the maintenance of animal life, the management of the spaces set aside to hold them transformed in line with shifting understandings of animal bodies over the course of the nineteenth and early twentieth centuries. Experiments in animal husbandry, including modes of animal management such as environmental enrichment (though the term "environmental enrichment" was only created in the later decades of the twentieth century), and the need to respond to increasing understandings of captive creatures had important implications for the provision of living space. As we know, keeping animals alive in captivity was an ongoing challenge, and the changing nature of the animal houses at the Zoo reflects a process of informal knowledge accumulation and a developing comprehension of animal bodies. Of course, by modern standards nineteenth- and early twentieth-century provision for the effective maintenance of animals in zoos was poor, but this was the result of an embryonic understanding of the conditions that could promote animal vitality. Previous captive spaces, such as the Tower Menagerie at the Tower of London, or the plethora of traveling menageries that toured the country did not configure animal spaces according to any particular kind of scientific rationale. This was where the modern zoological garden differed from its forebears. The perpetual process of experimentation upon animal bodies in the context of enclosure provision is clear not just at Bristol, but also throughout British zoological gardens where natural science was a founding principle.[26] Early on, warmth in the winter's cold, ventilation of cramped animal spaces, and the removal of dampness were considered to be of particular importance in the maintenance of health. The Society declared that "in the erection of houses for the accommodation of animals, the Committee have departed in some degree from the ordinary rules observed in this respect, by endeavoring to avoid the objections which, in the form of damp, cold, bad ventilation, irregular heating, excess of light," were characteristic of other animal accommodations of the time.[27] In December 1836, Honorary Secretary Dr. Henry Riley reported that the temperature of the Beast House ranged from 50 to 55 degrees Fahrenheit. This was probably 5–10 degrees too cool, he noted, stipulating that the health of the animals could only be preserved through the provision of warmer living quarters. Thermometers were installed so that temperatures could be properly monitored and regulated.[28] The provision of fresh air was also increasingly seen as essential to the maintenance of animal health, just as it was for human health at the time. Fortunately, the 1853 provision of increased ventilation in the monkey enclosure had the "anticipated" effect of reducing mortality considerably. It was noted some years after the erection of the 1902 outdoor monkey cage, too, that "it is believed that in the past the monkeys have been kept in too confined a place and without a sufficient supply of fresh air . . . following the idea adopted by the medical profession in their treatment of consumption known as the 'open-air treatment' an outside cage had been constructed."[29]

As we might expect, these kinds of experimental practices were especially important when it came to preserving the health of the Society's most valuable creatures. On the arrival of a new lioness in 1852, shareholder William Wheeler Green stated that "in his opinion true . . . economy would be best shown by making every effort to keep . . . such valuable animals (and of course the animals generally) healthy." He requested further amendments to the roof, "or at all events . . . to be allowed to move those stinking cats which just under the noses of the Lions are poisoning the air of the house," spreading disease in the process. His bracketed recognition of the need to keep all animals healthy does not detract from the special efforts afforded to the accommodation of the lion—as we have already seen—as a popular, prestigious, and valuable species. Similarly, an 1872 critique of the elephant house by the Zoological Society of London's superintendent, Abraham Dee Bartlett, led to the improvement of the enclosure, not least through the eradication of rats that, it turned out, had been feasting on the elephant's feet, in order to better protect "this valuable animal."[30]

Managing living space was not just about keeping valuable animals alive, of course. Some nineteenth-century enclosures contained elements of enrichment, though by modern standards these were very basic indeed. Prior to the later years of the twentieth century, what we now term "environmental enrichment" was usually based on trial and error, a "hope for the best" ethos, and was rarely informed by scientific data. While enrichment is a modern term, I use it loosely here in order to denote the ways in which captive environments were made into more than mere exhibition cases for the sentient beings within.[31] Simplistic enrichment might be identified in some of the first enclosures at the Zoo, where rocks and lumps of wood provided places for animals to rest in an otherwise bare enclosure (fig. 18).[32]

Effective environmental enrichment like this nods to an underlying though, I think, nascent interest in animal behavior as well as animal form, which was seen as necessary to the promotion of successful breeding. Indeed, reproduction was often understood as a signifier of happiness, or contentedness, in captive animals. It is now known, however, that exhibitions of hypersexual behaviors can often be indicative of psychological disturbance.[33] In 1854, it was reported that "your Committee has built a breeding den for

FIGURE 18. John Dilwyn Llewelyn, Griffon vulture, c. 1853. (Courtesy of Amgueddfa Genedlaethol Cymru/National Museum of Wales)

the Lioness, into which she and her young retire for warmth and quiet and has constructed, under this den, a boiler for the purpose of warming." Space, warmth, and seclusion were provided in order to encourage the successful rearing of young, and such practices were made in order to facilitate birthing across the years that followed.[34]

The provision of food was another key part of the great experiment. By providing the right foods and thus the right nutrition, animal bodies could be nourished and thus, in theory at least, effectively governed. By our standards, however, animal diets across the nineteenth and early twentieth centuries look decidedly peculiar. Captive beasts were routinely fed boiled potatoes, boiled rice, bread and milk in honey, boiled tripe, biscuits, pearl barley (like wheat but not especially nutritious), coffee, and apples stewed in sugar.[35] Feeding also entwined with breeding practice in the use of bitches to nurse big cat cubs when they failed to nurse from their mothers. In 1870, dogs' milk was supplemented with beef jelly, sheep's brain, and minced beef soaked in water, though the keeper drew the line at allowing the cubs to consume dog vomit. This combination was supposed to strengthen and invigorate the young animals.[36]

As well as this, however, the objectification of animals as individual specimens that could be zoologically understood and thus governed is linked to a parallel tendency to transform them and, subsequently, their species, into objects emblematic of the skills of those responsible for their governance. Individual young animals were not *only* lucrative commercial assets in the generation of vast crowds, as we have already noted. The Society and its keepers also saw them as testament to the human skill involved in the mastery of captive conditions.[37] Scientific ways of imagining animals were often complicated by this sense of human achievement. The first individuals of species never before reared in the Zoo, in Britain, Europe, or the world were heralded, and attention called to the great zoological achievements of the Society. Successful breeding implied mastery of the management of captive life, and, indeed, the Zoo was the site of a succession of breeding achievements across the nineteenth and twentieth centuries. Naturalist Frank Buckland noted in 1857 that it was curious how animals seemed to him to generally breed better at Bristol than they did at the nation's zoo, in London.[38] Sometimes patience was required as the Society was forced to wait for their experiments to bear fruit, however. In 1866, a pair of tigers were acquired. The committee noted that "no instance has been known of these animals breeding in England, but they have now an opportunity of trying the experiment." Two years later a litter of four Bengal tiger cubs was born in the Gardens, though London Zoo had beaten them to the post. If any of them had survived, it would have been the first time that tiger cubs had been whelped outside of London Zoo. While this and a further litter died that year, the Society achieved its aim of finally having whelped a litter of this prestigious species in 1870.[39]

LOOKING AT BEHAVIOR

Management, disease prevention, and care; all three derived from an increasing knowledge of animal secrets, and all three combined to enhance the likelihood of captive survival. This was a kind of human-animal relationship in which captive beasts were observed and interrogated in search of the answers to anthropogenic questions about the natural world.

Over time, the deepening of zoological knowledge combined with evolving medical and veterinary innovation and provision to better govern captive animals. This intensification eventually resulted in a shift in personnel. From the late 1920s, in accordance with Secretary Dr. Richard Clarke's plans to enhance the scientific credibility of the Society, a range of honorary positions was created. Medical professionals, many of whom had developed an interest in comparative medicine, occupied these positions. Such practices spanned across the muddy waters of the human-animal borderlands. The permanent appointment of these men contrasts with the informal and rather fleeting role of Webb in the 1840s and demonstrates the increased systemization and medicalization of animal management in response to a deepening comprehension of their bodies. A century after Webb, in April 1944, for instance, T. F. Hewer was appointed honorary pathologist and was responsible for the identification of disease, L. H. Kettlewell was appointed honorary dental surgeon, G. E. Henson honorary veterinary surgeon, and L. Harrison Matthews honorary biologist. While these men were primarily concerned with their day jobs, some at the University of Bristol, they comprised, in effect, a formalized group of professional medical men charged with using their highly specialized, albeit often anthropocentric (for most of these men were primarily concerned with human health), knowledge to preserve the vitality of the animal collection.[40]

These were not the first honorary positions, though. Samuel Stutchbury was the Society's honorary zoologist in 1836, but his role was more aligned with the acquisition than the treatment of captive animals.[41] The honorary positions of the mid-twentieth century did not come with a demand for full-time attendance to sick animals. Hewer, for instance, was only needed at the Zoo when an animal died, or when one had become seriously ill. In this regard, veterinary work remained reactive rather than preventative, piecemeal rather than comprehensive. Yet, the work of these men was complemented by an increase in written material, which collated disparate strands of zoological knowledge. Swiss biologist Heini Hediger noted that, prior to his 1950 publication *Wild Animals in Captivity,* there was minimal written guidance about the effective management of exotic creatures in captivity.[42] Nevertheless, a modest natural history library had been instituted at the Zoo from the 1930s in order to aid the surveillance and physical manipulation of animal bodies by providing easy access to an increasing body of substantiated zoological information and practical methodologies whenever they might be required.[43]

By the mid-1950s, the shelves of such libraries typically contained exhaustive digests on exotic veterinary medicine such as Halloran's 1953 "A Bibliography of References to Diseases of Wild Mammals and Birds."[44]

A significant turning point in the complexions and configurations of ways of thinking about and governing animal bodies was reached by the decades spanning the middle years of the twentieth century, and this affected *how* animals were observed. Deepening understandings of animal bodies and a somewhat transformed frame within which animals and animality were construed in light of the eventual digestion of Darwin's revolutionary thesis (for it had actually taken decades, possibly even into the 1920s, for his ideas to permeate society) had altered scientific views of animal life.[45] From the 1930s, but especially after the Second World War, the increasingly influential sciences of behaviorism and ethology heavily prejudiced the ways in which animals were understood as specimens, not only at the Zoo but also across the scientific arena. Concerned with the study of animal behavior, in the case of ethology usually in the context of a wild (i.e., nondomesticated) state, these approaches represent an increased scientific focus on the relationships between form, behavior, and environment. Heavily influenced by the works of Darwin, who showed that behavior as well as form could be inherited, ethology sprang to life in the 1930s through the work of biologists such as Julian Huxley, Niko Tinbergen, and Konrad Lorenz. In postwar Britain, the study of ethology was promoted by a diversity of natural history organizations all over the country.[46] Ethological study was an important step in formalizing zoological examination and knowledge of animal form and external behavior, and in the process, it entrenched the conceptualization of individual animals as predictable exemplars of species characteristics. Guidebooks reflect this through an increasing depiction of species according to geographical range, feeding habits, reproduction, and gestation cycles. The 1948 guidebook, for example, related the science behind reptilian cold blood, breeding behaviors, and the physiological mechanisms enabling the injection of venom. Such representations are quite removed from the surface-level depictions of earlier years in which morphology and superstition reigned supreme. People such as George Schaller and Dian Fossey, for instance, furthered these kinds of understandings through extensive fieldwork. Both worked with gorillas in the field in central/west Africa, significantly enriching knowledge of the species outside of Africa between the 1950s and early 1970s.[47]

This knowledge was then collated, allowing for a deeper understanding of the behavior of specimens in captive contexts. By 1973 the official zoo guidebook resembled an instruction manual, replete with comprehensive analyses of bodies and behaviors.[48] Importantly, ethology paid minimal attention to the realities of nonhuman consciousness (see a further discussion in chapter 4). Instead, this way of looking, of interrogating and knowing, was almost exclusively about the relationship between structure and external function. By the 1950s, the established centrality of behaviorism and ethology, and the fact that bacteria had eventually been accepted as the causative agent of a myriad of diseases among captive animals, changed the ways keepers looked at the animals under their charge.[49] This was clearly reflected in the recording of details in daily log books that went well beyond those relating to births, fatalities, and acquisitions that so dominated animal records in the nineteenth and early twentieth centuries. Compiled on a daily basis they contained observations pertaining to illness and associated physical and behavioral pathologies. The report compiled for 19 March 1953, for instance, documented that a jaguar had not fed the previous day, and so feces samples had been sent to the University of Bristol's veterinarians at Langford to test for infection. The practice of recording observations and responses regarding health and mortality developed quickly, culminating in the creation of a designated medical and veterinary committee in 1961, which, perhaps not coincidentally, overlapped with the 1961 formation of the British Veterinary Zoological Society, and which was devoted to such matters. By the 1970s, daily animal reports recorded an even greater amount of behavioral and pathological detail. The report compiled for 13 October 1970 logged precisely what the baby chimpanzee had eaten the previous day, how many times the great apes had been seen mating, and the consistency of orangutan stools, as well as a range of behaviors observed among other creatures.[50] Indeed, this all represents a move away from (not only) reactive veterinary medicine toward the development of a machinery of surveillance designed to keep disease, and death, at bay.

Commensurate with this intensifying understanding and surveillance of animal bodies and the percolation of the insights of Germ Theory from the middle of the nineteenth century onward, immunization against disease became progressively more typical over the course of the twentieth century. By 1934, lion cubs were being routinely inoculated against dog distemper. Much later in the century, in 1975, all cats received vaccinations for feline enteritis, and all juvenile primates were protected against tuberculosis, a highly infectious zoonotic disease.[51] Fecal analysis was also introduced and became increasingly sophisticated. Indeed, it was a vital set of procedures since feces were often the only animal "bits" available to veterinarians prior to the easy availability of tranquilizers.[52] By the 1970s, such analyses were routinely carried out at the University of Bristol's veterinary school at Langford. The examinations allowed for the detection of increasing varieties of pathogens and parasites in diseased animals.[53] The introduction of antibiotics and the so-called sulpha group of antibacterial drugs added an additional line of defense.[54]

In addition, postmortem examinations became increasingly detailed across the century, and this allowed for the implementation of a greater array of preventative measures. In 1928, the University of Bristol's Departments of Pathology and Cardiac Research conducted tests that concluded that a number of monkeys had died from lobar pneumonia and bronco-pneumonia. The Society was thus able to take steps to prevent the spread of infection to other animals.[55] Later, in 1968–69, the Society commissioned an extensive investigation to examine a virulent outbreak of *Pasteurella Pseudotuberculosis* among its captives. The investigators recorded occurrence of the disease, its mode of transmission, animal reservoirs, and possible methods of control. Mice were discerned to be the symptomless vectors of the disease, and they were exterminated in large numbers in order to eradicate them from the animal enclosures.[56] The formation of these kinds of committees and the execution of extensive investigations contributed to a mounting paper trail, which could be used to govern animals in a way that was—to a far lesser degree than in the nineteenth century—based on experimentation and more on the replication of already-tested methods.

Animal autopsies also increasingly served the purpose of facilitating the ever-important breeding of rare and valuable species. Deep within expired animal bodies lay the secrets of successful propagation. Indeed, this was increasingly important in light of the practical problems faced by Zoos in the post–Second World War era (see chapter 1). In November 1963, for instance, a male okapi calf was born at Bristol, the first since the elusive West African species had arrived in 1961. The animal died nineteen days later, with the postmortem examination revealing the presence of pulmonary aspergillosis and a secondary mucor infection. Further investigation showed that the clover hay in the okapi enclosure was heavily laden with organisms associated with the disease. A second male calf was born in April 1965, which also died, twenty-four days later. The postmortem revealed that this animal had died from a similar infection. For the third pregnancy, the findings of the postmortems were put to good use; as the female neared full term, she was moved to a grassed enclosure at Langford Veterinary School, and a substitute foodstuff was provided in place of clover hay. A calf was born in July 1966, which survived, being the first specimen of the okapi to be successfully reared in the UK. This individual formed the beginning of what later became a substantial breeding herd maintained by the Society at its Hollywood Tower Estate.[57]

Thus, the secrets held by dead animals and their excretions had increasingly become the keys to the ever more effective management of the collection. Beyond that, though, lifeless animal bodies continued to have the capacity to contribute to the accumulation of zoological knowledge more broadly. Indeed, a designated research collection, in large part composed of animal specimens formerly of the Zoo, was established at the Bristol Museum in 1931. The more popular exhibits of the natural history collection were no longer deemed conducive to rigorous zoological investigations.[58] Perhaps more tangibly, Dr. R. H. Johnson of the Bristol School of Veterinary Medicine developed a vaccine to combat feline enteritis in 1968, doing so through isolation of the causative pathogen in the body of one of the Society's deceased leopards.[59] In all of these contexts, death was the seed of life. The Zoo itself was the site of a deep and vibrant mortal metabolism.

Enclosure provision also reflected this combination of increasing understandings of animal bodies and behaviors and evolving medical and associated technologies. While the regulation of body temperatures continued to be understood as vital to the maintenance of animal health over the course of the twentieth century, stringent hygiene emerged as a priority after the ascendance of bacteriology.[60] It was enforced through the intensive sterilization of animal enclosures. While the living spaces of some animals were scrubbed with list lime and whitewashed from the earliest years at the Zoo, a greater emphasis on enclosure hygiene took hold in the twentieth century as bacteria, rather than foul vapors, loitering in captive environments were discovered to be the root of some of the most common afflictions among captive creatures. Indeed, hygiene was a common concern among zoos at the time; in 1911, Peter Chalmers-Mitchell of the Zoological Society of London wrote to Bristol's superintendent, H. Reginald Woodward, advising that the monkeys at London Zoo were kept alive through the use of "cleansing soaps," like Sapon, as well as formalin vapor candles designed to decontaminate the air of foul gases.[61] Post–Second World War, however, brightly lit, tiled enclosures became fashionable as the most effective means of facilitating the sterilization of living space. This "bathroom style" was considered most conducive to the maintenance of animal health through the easy elimination of dangerous pathogens.[62]

Big cats, for instance, benefited from the increasingly efficient elimination of pathogens. The outside cages in the Lion House had previously been floored with concrete, which was not only difficult to clean but also took an appreciable time to dry after rainfall. These floors were thus covered with asphalt in 1954, and this proved much easier to sterilize.[63] In this kind of enclosure schema, practicality took precedence over the provision of enrichment through naturalistic enclosure designs, which might allow room for pathogens to thrive.

The insights of behavioral studies, many of which, like Schaller's and Fossey's, had been rooted in the wild places, also illustrated the importance of crafting healthy captive groups in the perpetuation of natural behaviors; Hediger noted in 1950 that some animals naturally lived in social groups rather than in isolation, and so he recommended that these species ought to be maintained in that manner in captivity. Accordingly, solitary or pairs of monkeys and gorillas were surrendered in favor of larger family groups for the first time in 1968, and by 1983, there were six breeding groups in the new monkey enclosure and a group of nine gorillas comprising a family.[64] Groupings like this allowed the Society to promote healthy behavior by replicating social conditions prevalent in the wild state. Similarly, the management of breeding behavior was affected by the combination of ethological insights and the availability of changing technology. Surveillance technologies, employed by the Society beginning in the 1960s, minimized the level of human interference in the natural processes of animal births and early life. Simultaneously, they permitted an unprecedented degree of access to creaturely lives. Infrared and surveillance cameras were installed in order to observe the birth of the Society's okapis in the 1960s and were also deployed following the death of one of the Society's white tigers in 1967. They remained in the tiger enclosure until 1970 so that the expectant mother might be monitored at all times.[65]

Lastly, a deepening knowledge of animal bodies and behaviors as well as a heightening realization in the interwar years that there was a direct correlation between diet and mortality rates gradually led to changes in the types of experiments performed in the provision of nutrition. By the late 1930s gorilla diets, for instance, consisted of milk, fruit (bananas), vegetables, marmite, and stout. The provision of vitamins was an important aspect of these diets: both marmite and stout certainly contain good amounts of B vitamins—vital components in the healthy operation of the neurological and circulatory systems. This complemented the administration of vitamin D via cod liver oil, which had been determined to be effective in remedying deficient exposure to ultraviolet light through confinement.[66] All of this was part of the provision of a diet that could provide all of the nutrients needed for healthy bodies and minds, and it remained largely unchanged across the middle decades of the century.[67]

As in earlier years, however, experimentation in order to get animal husbandry and captive conditions "right" was not only about securing the vitality of animal life for commercial and enlightened scientific reasons. Breeding successes, for example, were particularly esteemed as exemplars of the Society's skill. Much publicity surrounded the 1934 birth of Adam, the first chimpanzee born in captivity in Britain, and the Society's successful rearing of the first female chimpanzee, Helen, followed in 1944. Later, in 1953, the Society noted with satisfaction that the squirrel monkey had bred for the first time in a British zoo and possibly in a global zoological collection. It went on to successfully breed black rhinoceros (1958), okapi (1966), and white tiger (1968) for the first time in any British zoo. On a deeper level, individual animals were also emblematic of the achievements of particular people. When an Indian and an African python were hybridized in 1971, keeper Alf Elliott proclaimed that he was especially proud of his successful breeding experiment. That year also saw the birth of Daniel, the first gorilla to be born—and subsequently survive—in captivity in the UK. Senior keeper Mike Colbourne admitted that he was "extremely proud" of the birth, since he had aspired to be the first person to successfully breed a gorilla in captivity in the UK.[68] For him, Daniel was emblematic of that personal accomplishment in animal husbandry.

Thus, the decades spanning the middle of the twentieth century witnessed both a deepening of zoological knowledge as experimental practices increasingly bore fruit, a reorientation in favor of promoting what were understood to be healthy behaviors, and an increasing array of medical and veterinary armaments, so many of them inspired by the revolutionary insights of bacteriology. This combination reflected an enduring consideration of animals as objects that could be known and then managed. These advances did not, however, signal the end of the "great experiment" of captivity, but rather an evolution that then continued into the conservation age.

EXPERIMENTATION AND GOVERNANCE IN THE CONSERVATION AGE

The decades spanning the turn of the twenty-first century witnessed a range of transformations and continuities in ways of looking at and managing animal life at the Zoo. In many ways, this reflects the relatively lesser prominence of ethology and behaviorism as disciplinary paradigms and the increasing professionalization of the sciences of conservation and ecology. Nonetheless, animals remained precious objects that needed rigid governance. Surveillance and intervention continued to be, and in fact arguably became increasingly, vital, and just as they had across the rest of the period, animals remained powerful tools of enlightenment. Guidebooks published in the period reflect the perpetual objectification of animal life by extending the tradition of positioning individual captive creatures as representatives of whole species. Indeed, objectification within a conservation schema came to be less about animals themselves and more about their distinctively industrialized relationships with people. Scientific objectification in the late twentieth and early twenty-first centuries is therefore not only about the animal as specimen, but also, and critically, about its *tenuous* place in a multispecies world. Consequently, recent guidebooks are less concerned with the dense ethological description prevalent in earlier publications and more with effectively conveying messages rooted in conservationism. In these publications, individual animal specimens served the primary purpose of embodying the fragility of whole species and whole ecosystems. In effect, every animal in every recent guidebook served not only as a flagship specimen, but also as an emblem of departure. The 1987 guidebook framed its subjects by warning that "the wildlife of our world is trickling like sand through our fingers." The 2001 guidebook similarly directed the visitor to gaze toward the critically endangered Western lowland gorillas on display at the same time as it warned that lemurs "are being threatened by slash and burn agriculture, charcoal production and mining for sapphires and other gemstones and minerals" in their Madagascan habitats.[69] In some ways, this returns us to the kinds of scientific looking practiced in the nineteenth century, where species were considered according to anthropocentric contexts.

The inscription of ecological fragility onto animal bodies added to the necessity of success in the great experiment that was management in captivity. The depth of knowledge acquired about the bodies of captive animals has been unprecedented over the past few decades. Husbandry practices shifted and developed in line with the emerging desire to protect brittle animals who hold within their fleshy, bony bodies precious genetic material that is vital to the propagation of species. The system of appointing medical and veterinary professionals to part-time honorary positions continued unabated until the latter part of the twentieth century, when it was considered necessary to introduce round-the-clock specialist veterinary care on-site. In 1981, a full-time veterinarian was appointed for the first time. By 2000 veterinary care at the Zoo was extensive; some 760 cases were attended to that year, covering 104 species.[70] Keepers also became experts in the care of specific animal groups. Together, keepers and vets formed a formidable defensive force against the powers of destruction wielded by disease and poor nutrition, in particular. Keepers and vets today are specialist scientists, and thus the level of knowledge of animal bodies and the variety of interventions available is quite extraordinary. The vet's reports from 1990 and 2000, for instance, reveal that the mechanisms of individual animal bodies were increasingly known and maintained through x-ray, electrocardiogram, blood sampling, and fecal analysis, all of which were made easier through the increased availability of tranquilizers.[71] The Animal Records Keeping System (ARKS) and the Medical Animal Records Keeping System (medARKS) serve the allied purpose of maintaining extensive records about animal behavior and health. This kind of knowledge and the machinery designed to extract it is vital, for animal husbandry remains experimental. Knowledge about the lives of captive creatures has been continuously accumulated over the course of nearly two centuries, and this knowledge has been and remains of use in the governance of captive creatures to promote healthy bodies, behaviors, and, indeed, species.[72]

In multifarious ways, captive animals were seen to hold in their bones not only the secret to effective management in captivity but also the secrets of effective preservation in the wild places. With the formation of the Bristol

Conservation and Science Foundation in 2006, animals at the Zoo have been subjects of a torrent of academic explorations, including recent studies relating to reproductive health in gorillas and breeding management of Livingstone's fruit bats.[73] The bodies of the dead also have an increasingly expansive role to play within a scientific schema dominated not only by the continuing desire to understand the workings of animal biology but also by the need to conserve species in their various environments. In 1992, adipose tissue samples were taken from two dead polar bears (whom we encounter in all their distressing glory in chapter 5) when they were dissected at the veterinary school in Langford. The tissues, together with further wild and captive examples, were used to investigate "the origin and natural function of massive superficial adipose tissue in wild polar bears" and "the distribution and site-specific properties of adipose tissue in wild polar bears."[74]

Today the Society has arrangements regarding the disposal of its animal corpses with research collections such as National Museums Scotland. Specimens are sent there to be held and then used in an array of research activities. The skeleton of the Society's last Asian elephant, Wendy, for instance, forms part of that collection, alongside at least 250 additional specimens acquired from the Society over recent decades. Many of these have been the subject of intense examination in specific scientific investigations, which have important implications for the understanding, and management, of species around the world. A study published in 2002, for instance, utilized the bodies of siberut macaques. One had been exhibited at London Zoo until 1933, and another had died at Bristol Zoo in 1997. The siberut macaque is found on Palau Siberu, Southeast Asia, and only two specimens, those examined in the study, have been cataloged to date. The study of their bodies found that the siberut macaque is a distinct species rather than conspecific with the Pagai island macaque. The disclosure of this differentiation expands understandings of the species and has important implications for the management of wild populations. Another paper, published in 2012, used the femur of a white-tailed eagle, which had been acquired from the Zoo in May 1968, in a study relating to skeletal stress.[75] Scientifically valuable, these fragments of stilled bodies were—and many remain—capable of deepening understandings of species in life, both captive and wild.

An increasing understanding of the association between species and environment, in many ways born out of the field studies that had intensified from the 1950s, and which grew in sophistication over the course of the second half of the century, led to an array of emerging enclosure practices that sought to promote changing notions of natural behaviors.[76] Enriching animal spaces could promote behaviors such as those displayed in a wild state, and the replication of wildness and wild places (or spaces serving as proxies for natural environments) has increasingly been construed as vital in the provision of animal welfare.[77] Indeed, stark and easily sanitized animal spaces gave way to more naturalistic environments, partly because of the changing sensibilities that we have already encountered over the course of this book, but also because more effective immunizations and antibiotic treatment minimized the risk associated with exposure to natural materials. While earlier enclosures were perhaps enriched in a rudimentary way, only in the last decades of the twentieth century did this become a practice in and of itself. In the late twentieth and early twenty-first centuries, enrichment was achieved through a combination of techniques. The Society's Gardens team, for instance, worked with the animal department to provide enclosure vegetation similar to that with which the animals might interact in the wild, though this is often somewhat tempered by the need to avoid poisonous plants and those that would not survive a British winter.[78] Beyond the provision of vegetation, the 1997 Gorilla Island exhibit provided private spaces where the animals could retire from public view, and presented food in ways that encouraged the animals' foraging behaviors. Features such as running water and an environment that varied seasonally were also designed to mentally stimulate the captive creatures, improving general well-being by replicating an ever-transforming habitat.[79] Indeed, food itself forms part of these enriched habitats. Red pandas, for instance, receive bamboo: very different from the limp boiled vegetables and mangled brains of earlier years.[80] Importantly, these facets of today's enclosures point to a rising recognition of the *mental* worlds of animals. Governance today is directed toward both brittle body and fragile mind.

These kinds of practices equally contribute to the propagation of precious endangered species: The AmphiPod exhibit (2010) provides the precise environmental conditions—a "love shack"—required to promote breeding behaviors among some of the world's most endangered amphibians, the lemur leaf frog and the golden mantella. The facility also isolates them from the threat of disease, primarily of a fungal nature, that is decimating wild populations.[81] Even in the era of conservation breeding, however, there remains a strong sense of institutional accomplishment attached to the successful breeding of endangered species. The first breeding of Savu pythons and black marsh turtles, and the fact that Bristol Zoo was the only zoo in Europe to be breeding the giant Asian leech, have all been sources of institutional satisfaction since the turn of the twenty-first century.[82]

CONCLUSION

Considering animals as part of a "great experiment" spanning across nearly two hundred years, but changing in complexion along the way, has formed another key layer of human-animal relationship at the Zoo. It lay alongside considerations of captive beasts as precious, multisensory objects of amusement and instruction. To say that the process of maintaining captive creatures in an artificial semi-urban environment was a "great experiment" across the period is no exaggeration. Zoological societies, their staff, and the veterinary and medical professions simply did not possess deep knowledge of animal bodies at the outset. Therefore, from the beginning, a critical layer of human-animal relationship was about trying to understand, often by trial and error, how precious animals might be kept alive and healthy. Today, the "great experiment" is no less fundamental to interspecies relations at the Zoo, though of course the scientific ethos within which that is taking place has transformed into one largely preoccupied with conservation and the maintenance and regeneration of biodiversity. As we have already seen, captive animals were considered precious across the period, and this meant that the desire to maintain animals in consequence of their value endured.

The secrets held deep in animal bodies were not merely for use by the Society. Its primary aim was to facilitate effective governance of captive creatures, but it was also to better educate visitors to the Zoo and to contribute to understandings of animal (and sometimes human) life. Moreover, it becomes clear that relating to animals in terms of the knowledge that can be gleaned from their precious bodies did not rest on scientific intention alone. This layer of human-animal relationship was in many ways dependent on changing technologies and changing scientific knowledge within and beyond the discipline of the animal sciences. Changes in medical treatments, the rise of bacteriology, and technologies that permitted deeper explorations of animal bodies all allowed for the perpetuation and the deepening of a human relationship with animals founded on the extraction of knowledge.

But this was a layer of relationship that, until very recently, was founded on a kind of knowledge of bodies and behaviors that framed animals as fundamentally predictable; indeed, almost mechanical. These animals were objects of understanding, objects of deep and profound knowledge. And yet animals are by any account much more than their physical selves, and they are more than just constituent parts of a homogenized species group. They have lives that are independent and individual, and many of them experience the world in ways that chime with our own experiences as conscious subjects-of-a-life. These inner worlds, these individual experiences of the world, form the basis of the layer of human-animal relationship to which we now turn.

4

Minding the Human-Animal Borderlands

O scar, a Sumatran orangutan, was born at Bristol Zoo in May 1971. Having been abandoned by his mother, Ann, senior keeper Mike Colbourne became his surrogate parent. He fed him, changed his diapers, and rocked him to sleep. Over time, intensive hand rearing led to Colbourne developing a deep affection for the young primate. In fact, he developed similar attachments to a number of the great apes he was charged with keeping during his two decades or so working at Bristol Zoo. However, Colbourne's approach to the animals in his charge was complicated. He studied their bodies as repositories of zoological knowledge *at the same time* as he was subject to a deep emotional attachment (fig. 19).[1]

Not only did he see his captive wards as objects from which knowledge could be extracted, but he also saw them as individuals with complex personalities. This emotion, this recognition of personhood, of the presence of conscious beings that are more than objects of desire, amusement, or instruction, forms another layer of human-animal relationship in the Zoo and beyond. The creatures we invite to be parts of our families, those who talk to us through our television and smartphone screens, and those we demonize, too; they all reflect the ways in which we *mind* animals in the world; the ways in which we frame animals in the context of emotion and shared worldly experiences. It is now almost beyond dispute that Descartes, and the many others who expanded upon on his theorizations, was mistaken; that

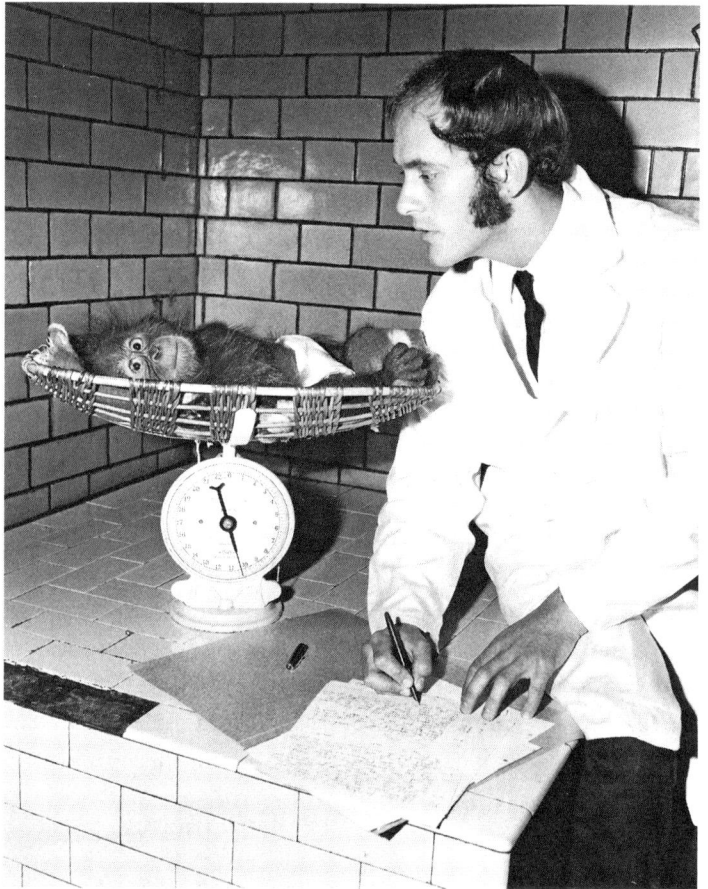

FIGURE 19. Mike Colbourne with Oscar the orangutan, 1971. (Bristol Zoological Society Archive, courtesy of the Bristol Zoological Society)

animals are without doubt thinking, feeling, sentient beings who experience the world in ways that exceed the merely physical. Behavioral and neuroscientific research provides overwhelming evidence in this regard.[2] Recognizing the reality of individual experience has increasingly led to the development of emotional attitudes toward animals, and sometimes to inappropriate anthropomorphization. But these are often part of the same phenomenon: the attempt to recognize an individual subject within the homogenous mass of their species.[3]

This recognition of individuality was complicated, however, not least by the fragile and fluid status of the human world as separate from the rest of nature. Oscar was often explicitly represented as a liminal beast who blurred the imaginary boundary separating man from animal. He was frequently depicted with a child's toy, and as a baby, he was often made to wear a diaper (for obvious practical purposes associated with the practice of hand-rearing). In 1973, a local newspaper reported that he was growing excited like a child at the imminent arrival of Christmas: "Well it is Christmas and you can't blame an orang-utan if he wants to get into the act. Take Oscar, the two-year old. He's as bright as a button, they'll tell you at Bristol Zoo. And that's the way Oscar intends to stay over the holidays, with a little help from his animal friends."[4] Anthropomorphic depictions like these symbolically domesticated the creature, positioning him unambiguously within a *human* moral circle. Primates are, in this regard, rather confusing to us. They are not human, and yet their skins and behaviors are inscribed with an essence of humanness. They occupy a liminal realm in which they are neither one thing nor another. Thus, amid Oscar's anthropomorphization and the recognition of his individuality, there was nonetheless a persistent awareness of the supposed absolute distinction between what was imagined to be the essential character of the young orangutan and the human world of his zookeeper family. There were concerns, for instance, that problems might arise if the youngster was not taught that he was indeed *not* human.[5] Such anxieties and ontological confusions pervade the human-animal borderlands, where the recognition of individuality and mind draws animals into human worlds and in the process disrupts the alterity attached to their forms through processes such as commodification, and their conversion into subjects of scientific exploration and spectacular encounter.

Thus, between his birth in 1971 and his departure for Chester Zoo in 1979 Oscar was a shape-shifter. He took multiple figurative forms that reflect the diversity of human-animal relationships enshrined in the institution of the zoo. As valuable commodity, object of the zoological gaze, and spectacular amusement, he was certainly representative of his species. Oscar was also, however, perceived and presented as a thinking, feeling, loving, fearing individual subject whose animal otherness was far from absolute. At the heart of human-animal relationships rooted in the provision of care, then, was a way of looking that conceived of captive creatures as subjects-of-a-life who were recognizable and with whom one could emotionally engage. *Similar but dif-*

ferent, his status was chimeric, and this allowed the human-animal relation-ships in which he was implicated to fluctuate. Similarity and difference have infused human-animal relations for thousands of years; indeed, that confu-sion arguably defines our dwelling with animals in the world.[6]

Oscar was very much of his time. The cultural context of the postwar de-cades played an important part in conditioning people to respond to ani-mals in ways that were increasingly infused with this kind of emotional at-tachment. In contrast, nineteenth-century representations of individuals and species were often rooted in a perception of the natural world as an obstacle in the way of the magnificent purposes of human civilization. Later, and espe-cially into the twentieth century, the Zoo was constructed as a place where in-terspecies kinship and affection prevailed, and this reflected not only chang-ing ways of looking at animals but also, of course, transformations in the Zoo's self-image. Personalities for zoo creatures were not only recognized but also intensively cultivated in newspaper columns, in the emerging medium of television, and also by the Society itself, who explicitly deployed anthro-pomorphic devices to enhance the popularity of its Zoo. In some ways, the twentieth-century zoo became a *home* for exotic creatures who were evermore immersed in a multispecies social fabric.[7] The intensity of this way of looking eventually faded, but by the end of the twentieth century an undercurrent—linked to the recognition of animal individuality—of powerful emotional at-tachment remained and was increasingly married to conservationist agendas.

This rather convenient story of change over time is not as straightforward as it seems. Not all animals were subject to these kinds of responses in the way that Oscar was. As we know, charismatic megafauna often formed the most popular animal attractions. It is no coincidence, then, that they were also the creatures whose individuality was recognized most readily over the course of the entire period. During the 1960s, zoologist, ethologist, and popular socio-biologist Desmond Morris surveyed a sample of eighteen thousand children. He asked them about their favorite zoo exhibits: 97.15 percent indicated a preference for mammals, while over a quarter of these children pointed par-ticularly to the chimpanzee (13.5 percent) or the monkey (13 percent). This popularity, he argued, was attributable to the species' possession of certain physical characteristics, many of which are akin to our own. Rounded out-lines, hair, an often-vertical posture, flat faces with familiar expressions, and the ability to manipulate small objects allowed for a perception of a fundamental kinship across species boundaries, and this encouraged both the recognition of individuality and a slippage into anthropomorphic constructions. In contrast, animals such as insects, reptiles, amphibians, and fish have far fewer features that we might construe as anthropoid. Because of this lack of perceived similarity, these kinds of animals have tended to be less popular, less easily transformed into objects of affection, and more often negatively interpreted. Humans do not, Morris argued, see "animals as animals, but merely as reflections of ourselves, and if the mirror distorts too badly we either bend it into shape or discard it."[8] For most people, a greater ontological chasm needs to be crossed in order that empathy develop toward creatures that appear more alien to our eyes. This fundamental attraction to similarity accounts for the "grades of difference"—a hierarchical speciesism—that manifested when zoo animals were imagined and presented as subjects over the entire span of the Zoo's history.[9] It is no surprise, then, that the animals that we will consider in this chapter were predominantly mammalian. The charismatic qualities of animals directly affected the ways in which they were construed. Zoo animals that became objects of recognition and affection tended to be large or charismatic creatures such as gorillas, elephants, seals, and monkeys, and these were the creatures about whom most noise was made, by the Society, its staff, and visitors alike. Nonmammalian groups were usually imagined quite differently. The focus in these contexts was less on recognition and similarity and more on the otherness of anatomies and behaviors, and this obscured animal individuality. In order to understand this, I refer throughout this chapter to two classes as representative examples of captive creatures whose individuality generally remained obscured in most circumstances: reptiles and fish.

This is not to say, of course, that stratified perceptions of both mammalian and nonmammalian animals were universal. It is entirely possible that some people did not relate to any kind of zoo animal in emotional terms. Equally, some interpretations of species with fewer anthropoid characteristics were both unambiguously anthropomorphic in tone and implicit of the recognition of individuality. The program for a 1932 Zoo carnival, in aid of Bristol's hospitals, for instance, featured an advertisement for the aquarium in which fish were depicted as thrilled at the prospect of meeting human visitors: "we have an ocean—a notion, pardon!—that the aquarium will yield a specially big catch in our search for your silver and gold. The fish will be delighted to see you, and if they could speak, would say (T)hanks (oi!). Pass in, please."

"Tame" boas and "talking" tortoises were later featured in 1960s press reports.[10]

This is important, for it tells us that the mind is a uniquely personal place where the world is often construed in highly individualized terms. Consequently, layers of relationships between human and animal can be shaped in tremendously narrow contexts, the complexion of the human-animal borderlands thus uniquely configured. W. M. Jones's humorous book *Jarge Balsh at Bristol Zoo* (1934) relates the adventures of one Jarge Balsh, a visitor from a small Somerset village, enjoying a day out at the Zoo with friends. Upon seeing an orangutan, Balsh remarks that it had "a horrible lookin' feace wi' long ginger whiskers an' long ginger hairs all over the body. I can't get the zight o' the thing out o' me mind, an' ivery time I da see Butcher Brown do bring it all back. Butcher is ginger an' he do sheave his upper lip an' chin, an' lef underneath his chin, like the vigger in the Jack-in-the-Box."[11] This might be a fictional encounter, but its sentiments ring true nonetheless. People understand living things within a frame of reference configured by their own cultures and individual experiences and concerns. The real-life journalist Max Barnes contextualized two mirror carp at the Zoo through a lens to which nobody else had access, firmly rooting his imagination of the animals within his own life-world: "My own two favorites are the magnificent mirror carp, Duke and Duchess. I like to watch them cruising majestically round, dwarfing even the large Golden carp with whom they live. My interest in the Duke and Duchess is partly sentimental. They are reminders of boyhood days spent fishing on the banks of Duchess' pond, Stapleton."[12] The upshot of all of this is that perceptions of, and feelings toward, animals are complicated. The story of animals and the perception of individuality, which I now relate, belies the fact that the worlds of our imagination are more chaotic and eclectic than we can possibly know.

FEAR OF THE ANIMAL EARTH?

Nineteenth-century interpretations of species were often underscored by an emphasis on the subordinate position animals were supposed to play in relation to humans in Western cultures. At the beginning of the modern period and commensurate with the philosophies underlining so many of the world's religions, nature and culture were separate realms.[13] Humans were ontologically divorced from the supposedly lesser, and primal, denizens of the natural world. Yet, they were clearly understood as having lives and agendas of their own, which were generally commensurate with their positions in relation to humanity. Indeed, these lives often unfolded in ways that were deemed to be detrimental to the grand designs of the masterful human being at a time when most people were generally more vulnerable to the violence of nature than they are today. Across the Victorian period, "bad" animals were proven dangers to either body or property. They were driven by passion, greed, and malice; they subverted the rightful order of nature in which man held sway over the beasts of the jungle. "Good" animals, meanwhile, were willing servants, many of whom were capable of receiving and displaying affection in relationship with people.[14] The Zoo's 1843 *Catalogue of the Animals* portrayed monkeys as vicious, conniving miscreants. It was said that they consciously committed "great depredations in the cultivated grounds near the forests." Mandrills were similarly portrayed as treacherous and ill-tempered. The guanaco, meanwhile, was decried as a creature frequently prone to "spiteful malice"; parrots supposedly had a "mania for destruction"; and the tiger had an unquenchable taste for cruelty. Nearly forty years later the nub of these conceptualizations remained; in the c. 1882 guidebook some species of monkey ruminated on mischievous matters while others were thieves and vagabonds, "waging war" on human societies.[15]

The perception of conflict between humans and the "bad" animal inhabitants of a largely antagonistic natural world provided the foundations of the recognition of animal character and its frequent conversion into anthropomorphic animal knowledge in the nineteenth century. This is especially clear in the Society's depictions of reptiles. Emphasis was placed on size, power, the gory particulars of reptilian predatory habits, the explicit threat posed to people in places where coexistence was inescapable, and their unbridled cruelty. Indeed, it was determined that "God" had created only a relatively small number of these creatures so that their destructive ways might not spread uncontrollably and threaten the extension of human hegemony.[16] In contrast, species that did not pose a threat to human mastery were subject to more favorable portrayals, which often cast them in the likeness of domesticated pets. Lemurs, for instance, were "entirely free from all the malice, mischief, and treacherous propensities of [monkeys]."[17] Depictions such as these recognized character among whole species in the context of the nature/culture binary and the chasm separating human beings from the beasts of the jungle.

In the midst of all of this, individual animals at the Zoo were also recognized as persons among the homogenized mass of their species. An elephant by the name of Sambo Elephas was described as "kind and docile" during his stay in the Gardens in 1842, for he was easily controlled by his keeper.[18] In addition, the very attribution of nonscientific human names to certain animals served to construct personhood. These creatures were, in effect, pets constituting part of a larger zoo family. Of course, awarding names to captive animals was not a particularly novel phenomenon; a bear named Old Martin and a lioness called Fanny were both displayed in London's Tower Menagerie in the late eighteenth and early nineteenth centuries.[19] Among the Zoo's animal somebodies were Jenny, a rhesus macaque who lived at the Zoo around 1853, as well as lions called General, Jupiter, Ajax, Hercules, Mercury, and Pegasus. Jenny was a typical name for primates of the time, while the unashamedly masculine names given to male lions endowed them with qualities intimately associated with anthropomorphized gods and human warriors.[20]

This recognition of animal individuality was at its richest and most emotional in the relationships between specific people and the animals with whom they were intimate companions. Sociologist Clinton Sanders posits that people with deep emotional attachment to their pets perceive them not as species or breed representatives, but as persons with distinctive characters. They are, according to Erica Fudge, "a pet first, an animal second."[21] This was clearly the case in the context of the relationships between keepers and animals kept at the Zoo and reveals that the recognition of individuality manifested on the ground and not only in the more surface-level contexts of literary representations and naming practices. This relationship, however, was complex. While the emotional dimensions of such relationships were not usually made explicit in the nineteenth century, the levels of intimacy in evidence point at least to the management of animals as individuals with their own specific needs. In some ways, the line between concern for animal welfare driven by the desire to preserve animals as both precious commodities and repositories of information, as seen in chapter 3, and husbandry motivated by emotional attachment cannot be wholly determined. Emotion and practicality overlapped and synergized in the care afforded to captive creatures. The relationships of care and governance are not, as Braverman has noted, mutually exclusive. In many circumstances, they are inextricably entwined, and this infantilizes animals, transforming them into childlike entities who are dependent on their authoritative superiors.[22] While Braverman considers Foucault's notion of "pastoral power" in the context of twentieth- and twenty-first-century conservation, it can also be applied to earlier periods where affection complicated the detached and objectifying desire to preserve and display valuable animal objects.

In 1841, the physical fragility of two chimpanzees required their close management by head keeper William Almond Green and, it seems, his wife. The male frequently breakfasted with the two of them, while Green first chewed the female's food, before he administered it to her by mouth. In similar fashion a keeper devotedly hand-reared newborn lion cubs in 1870. He washed and warmed the animals and supplied two bull-terrier bitches to nurse them. When the lion cubs failed to respond to these measures, he prepared a mixture of cream and "water soaked in minced beef" to supplement the milk, before he moved the animals on to solid foods.[23] By modern standards, these interactions took place at astonishingly close quarters. They reflect not only the informal—trial and error—character of exotic animal husbandry at the time, but also the intimate nature of the interspecific relationships that developed at the Zoo from its earliest days and point to the presence of intimate emotional relationships. At the heart of these relationships was an appreciation of the individual natures of the animals under the care of devoted keepers. Indeed, recognition of individual animal suffering was an important influence on the ways in which animals were managed. If it was considered that an individual's suffering had become intolerable, it would be put to death in a potent exhibition of mercy. A lion was destroyed in 1839, for instance, "he being in the last stage of disease [and] suffering much." An old, blind camel was similarly slain in 1884 and a lion with a spinal complaint euthanized the same year to end its misery.[24]

Guidebook descriptions, naming practices, and interactions in the context of animal husbandry were each reliant upon, and reflective of, the conceptualization of animals as more than mere objects of commercial, spectacular, and scientific value. These layers of relationships were complicated by the frequent recognition that animals were subjects with rich inner lives that were in some way accessible to human observers. Yet the reality is that these lives were fundamentally challenging to access. Nobody could really know the precise degree to which an animal suffered, or whether it was indeed happy with the lengths afforded to its care. The realities of emotional looking, then, rested upon what, precisely, one was conditioned to find in the face of the animal other, and this rendered the recognition of animal individuality rather plastic in character. In 1886 ethologist and psychologist Conway Lloyd Morgan thought that he recognized pride in the face of the Society's lioness as she withdrew her cub from his reach. His friend, animal artist John Trivett Nettleship, thought otherwise, seeing instead fear and unease. The two men read animal faces in contrary ways because their emotions and individual contexts influenced the ways in which they sensed the presence of a lively being. Indeed, Nettleship's perception of fear in the face of the animal other was not unusual. A "lover of animals" wrote to the *Clifton Chronicle and Directory* in 1885 commenting on the unhappiness of captive creatures and specifically pointing to the "poor brown bear shut up in a den at the bottom of the bear pit."[25]

Beyond pity, the shooting of catapults and the burning of skin with flaming cigarettes that we encountered among the spectacular zoo pastimes of the nineteenth century (chapter 2) reveal not only a consideration of animals as amusements but also a particular kind of response to the perception of animal sentience. Erica Fudge persuasively argues that in order to bask in cruelty "there must be a recognition of suffering, but such a recognition implies sameness."[26] Some—mainly juvenile—visitors to the Zoo, then, rooted their cruelty toward captive animals in the recognition of the presence of an individual capable of experiencing bodily and emotional sensation. One episode, reminiscent of the sentiments underpinning the gruesome duels of the medieval and early modern bear pit, shows that some visitors induced the spectacle of suffering and, in this case, death. In 1878, two bears were discovered lifeless in their enclosure. Upon inspection, it was thought that they had been poisoned with arsenic, which had likely been administered through feeding. A later report corrected this information, noting instead that they had died not at the hands of arsenic but of strychnine, a chemical that had been commonly used in the North American war on wolves. The poison is one of the most agonizing toxins known to science. It produces muscular convulsions, taking between two and three hours to kill.[27] It is at least possible that the use of strychnine was intended to induce in the bears physical suffering rather than instant expiry. The convulsions would have inflicted visible pain over a significant period of time.

Thus, the recognition of individuality and conscious experience, and its manifestation in affection or a desire for cruelty, was ubiquitous in the nineteenth-century zoo. Yet, this recognition was not evenly distributed among all species. More than any other class of creature, primates troubled the human-animal boundary in their position as liminal entities, so obviously animal and yet apparently so near to the human in appearance and behavior. Donna Haraway notes that "monkeys and apes have a privileged relation to nature and culture for Western people: simians occupy the border zones between those potent mythic poles."[28] At least since Aristotle, it was observed that, anatomically speaking, humans and other great apes were enormously alike. So obvious was the physical likeness that scientist and physician Edward Tyson named what was almost certainly a bonobo that he dissected in 1699 *Homo sylvestris,* meaning "wild man of the woods." Later, "Father of Taxonomy" Carolus Linnaeus, unable to discern sufficient anatomical distinction, categorized humans as well as their primate relatives as part of the same genus in his canonical *Systema Naturae* of 1735.[29] Yet anatomical resemblance did not equate to likeness of being. Positioned in a great *scala naturae,* a hierarchical chain of being in which all entities had their designated place, from God at the top to the minerals of the Earth at the bottom, humans were frequently recognized to be similar to, but nevertheless absolutely distinct from and superior to, the rest of the animal world—primates included. The publication of Charles Darwin's *On the Origin of Species* in 1859, with its implication of human descent from a primate ancestor, and his subsequent 1871 *The Descent of Man,* which made that implication explicit, challenged accepted notions of human biological distinction from the rest of the animal world. Darwin's 1872 *Expression of the Emotions in Man and Animals* did much to hasten the disintegration of the boundary separating the minds and emotions of humans and other animals. Despite his insights, however, a belief in some kind of fundamental human exceptionalism remained intact. At least since the Enlightenment, features such as speech, bipedal locomotion, and the possession of souls were held aloft as fundamental dissimilarities between humans and the brute creation. Vehement opponents of Darwin's thesis such as Richard Owen, and even supporters of Darwin's theory such as anatomist Thomas Huxley, continued to insist on a variety of characteristics that absolutely rendered the human being ontologically separate from the rest of nature.[30] Over the course of the nineteenth century, an increasingly tenuous certainty in human exceptionalism, troubled by the increasingly observable resemblance between primates and humans, positioned primates, and great apes in particular, at the nexus of the nature-culture/human-animal binary.

This meant that the recognition of these creatures as individuals, and the emotional responses that followed, were infused with anxieties about what it meant to be human. The only official publication for visitors surviving from the period prior to the publication of Darwin's theses is the 1843 *Catalogue of the Animals*, which confirms the perception of unsettling convergences between humans and the anthropoid apes. Chimpanzees were described as anatomically resembling humans, but the publication dismisses the oft-quoted notion that chimpanzees were simply humans in a lesser state of civilization. Instead, it insisted that humans could be effectively distinguished from primates through a number of anatomical, physiological, and spiritual differences, such as the ability to walk upright with their head facing toward the heavens (a theological distinction cited since classical times).[31] The animal was also, however, imagined as having a manner that, "however distant," approximated to "those of the young of our own race." In this way, both similarity and distinction were simultaneously implied. Furthermore, human expressions observed on the faces of monkeys were ridiculed as "ludicrous," confirming a perception and concurrent rejection of sameness in the mid-1840s.[32] The press and the public, meanwhile, perceived an uncertain likeness between themselves and the apes they saw on display. In 1841 it was reported that a male chimpanzee went "mad with grief" after the death of his female companion, its uncanny screams, moaning, and crying "like a Christian" disturbingly evoking those of a human being.[33]

The perception of similarity of being was easier, if sometimes more troubling, in relation to mammals than to more ontologically "Other" classes like reptiles and fish. Creatures that were further from the human condition were more frequently objectified as strange, fascinating, sometimes frightening curios in the nineteenth century. Fish were predominantly construed unsentimentally in light of their strange appearance rather than in response to any sense of ontological proximity. Until relatively recently, fish were habitually omitted from the category of creatures we term "wildlife"; their extensive use as sources of sustenance, in particular, rendered them marginal entities in human ideas about wild things.[34] In the nineteenth century, the Society referenced fish only as human property and as a source of human conflict with crocodiles around the watercourses of the Indian subcontinent, rather than as wildlife in their own right.[35] Further, the oceanic expanses of the world tended to evoke fear in response to a feeling of vulnerability in the face of an inhospitable and inaccessible vastness. Some of the first public aquariums depicted a correspondingly dark, shrouded, definitively "other" watery world.[36] Though they were far from unpopular exhibits, fish clearly were positioned and perceived at the Zoo less as recognizable individuals and more as striking oddities emanating from mysterious aquatic depths, and they were thus well outside of any humanist moral circle.[37]

Less definitively "Other" on the scale of difference, reptiles, like birds, were admired for their appearance rather than for their proximity to humans in body or mind. Where they were recognized as individuals, it was usually within a wholly negative frame of reference. Among a maligned group of animals, frequently associated with the primeval and often defined by their "destructive" natures, they were broadly disliked, certainly from the early modern period, and the threat posed by many reptiles was elevated in the earliest guidebooks. The guidebook appearing around 1882, for instance, recounts writer and author Valerius Maximus's (c. AD14–37) account of a 120-foot reptilian "monster" that was slain and the skin sent to Rome as a trophy. A further anecdote described the "largest snake ever known," which devoured beasts and men whole. Living in the lakes and marshes of the steamy Amazon rainforest, it was a creature of "frightful magnitude" and "deleterious nature," lurking in wait for its prey.[38] In spite or perhaps because of this, such creatures were consistently popular sources of titillation at the Zoo. Presumably, this was on account of the perceived threat they could pose to human vitality and the associated repulsion directed toward their violent forms. Indeed, Florence Grinfield asserted "what the Zoological Gardens would be without the snakes, a cake would be without the citron and currants." On rare occasions, ad hoc spectacles were crafted around the sense of danger emblazoned on reptilian forms. In 1893, Esme Edwards, the stepdaughter of Sir Greville-Smyth (a naturalist and animal collector who lived in a stately home close to the Zoo), caused a sensation when she entered the Zoo's snake pit. She proceeded to allow the creatures to slowly wrap themselves around her body, much to the perverse delight of onlookers.[39]

AT HOME WITH THE ZOO PET

As the nineteenth century gave way to the twentieth, the nature of this layer of human-animal relationship at the Zoo transformed increasingly rapidly. Negative readings of antagonistic animal character began to fade in light of the eventual dissemination of Darwin's insights and the slight dissolution of the human-animal boundary. Animal individuals were seen more and more in positive terms. Over the course of the first half of the twentieth century, they morphed into increasingly friendly, cuddly, and agreeable almost-people, and this way of presenting and perceiving animal life persevered well into the post–Second World War era. Indeed, by the end of the nineteenth century captive animals, including those formerly characterized as "bad," had generally come to be seen in terms that were more positive. Formerly treacherous and deceitful monkeys had morphed into "playful" "folk" or "fellows," while a growing number of creatures like the gazelle were described in light of their capacity to make good pets.[40] This positive turn in the comprehension of animal character had important consequences. By the early decades of the twentieth century, the Society had without doubt started down a path on which captive animals were construed not only in positive terms but also as civic pets belonging, in varying degrees, to the people of Clifton and of Bristol more broadly. A pamphlet for the summer 1919 Zoo Carnival, for instance, featured three monkeys and an elephant, Rajah. One of the monkeys is attired with a suit, hat, and bow tie, while another wears a coat and is painting the side of the elephant, who is puffing away on a pipe. Two of the monkeys are apparently using Rajah's trunk as a telephone system through which to communicate with one another (fig. 20).

Representations of affable zoo animals were increasingly used by the Society over the course of the pre–Second World War period, but especially from the 1930s, in order to attract and engage visitors. Influenced by the increasing ubiquity of pet keeping across classes—though pet keeping had been on the increase in Britain for some time—and the rising cultural presence of amiable cartoon animals such as those featured in films and television programs produced by the likes of Walt Disney, individual captive animals were routinely transformed into domesticated pets.[41] In a distinct shift in tone since the previous century, captive creatures were predominantly presented as peaceful rather than as the violent avatars of earlier periods. By the mid- to late 1930s, chimpanzees, for instance, were presented as playful, "naughty" children, while other creatures were decreed to be "loveable."[42] At the same time, the kinds of negative sentiments and negative anthropomorphizations that featured in the Society's nineteenth-century depictions of fierce creatures swiftly fell away. By 1937, only a handful of carnivorous animals such as the big cats, lynx, fox, and bears remained cast as "cruel," "cunning," and "treacherous" beasts.[43]

FIGURE 20. Front cover of program for the 1919 Zoo Carnival.
(Bristol Zoological Society Archive, courtesy of the Bristol
Zoological Society)

Negative portrayals aside, the Society deliberately began intensifying the
imagination of its animals as individuals actively seeking communion with
humans, and especially with children. The names the Society generally chose
for the animals increasingly rendered them more recognizable and approach-
able than their scientifically depicted or prestigiously named nineteenth-
century predecessors. Big cats became known as Minnie, Kitty, Puss, and

Susan, while males had names like John Bounder.[44] Newspapers were also exploited as means of promoting the domesticated charms of the Zoo's animals. The anthropomorphized adventures of the Society's sea lion, Dinkie, were serialized in the *Bristol Times and Mirror* in the mid-1920s, for instance.[45] A young chimpanzee, Timmy, was similarly anthropomorphized toward the end of the 1930s. In one article, he addressed the audience directly: "Hullo, I'm Timmy. What do they call you? I'm a 17 week orphan chimp at Bristol Zoo, and Mrs. Amy Jones, who is looking after me, says I'm old enough to meet the public at last . . . I'll be looking forward to meeting you. Cheerio for now!"[46]

The rise of what was, to all intents and purposes, the zoo pet—a being subject to an intense form of affectionate individual and civic looking—was clearer still in the ways in which captive animals began to be converted into mascots. Their individuality became intimately associated with anthropocentric— and increasingly domestic—concerns. Mascotism at the Zoo has its origins in the nineteenth century, but it developed into something more emotional in character over the course of the first half of the twentieth century: a pair of baboons became mascots of the 68th Battery of Royal Artillery and the 2nd Battalion Royal Fusiliers during the Second Boer War (1899–1902), while a Russian bear became an icon of the 11th Hussars in the mid-1920s.[47] In these contexts, anthropomorphic devices were deployed as a means of transforming the animal into an almost-person who was emblematic of the plight of men at war. One particular monkey, Jack, was used around 1916 to draw attention to human hardships on the Western Front. He supposedly served alongside British soldiers, and references to gassing and the specter of physical mutilation were used to encourage public donations so that cigarettes might be sent to the troops on the frontlines (fig. 21).

By recognizing individuality and then making particular animals relatable to people almost *as* people, it was possible to promote a sense of enduring relationship. As part of the increasing construction of the recognizable, however, animals were crafted so that they reflected accepted social norms. In assigning gendered roles to primates, for instance, both the Society and the press, corresponding with wider trends in which family life was increasingly promoted as a source of stability, positioned females rigidly as mothers and wives whose duty it was to care for their "children" and submit to their "man-the-hunter" husbands.[48] In so doing, primates were employed as surrogates for people, and a clear connection was drawn between the individuals of pri-

Mascot Monkey, Clifton Zoo. 1449.

CI...ETTE FUND
FOR MY
REGIMENT.

I HAVE BEEN MANY MONTHS
WITH THE TROOPS IN FRANCE.
HAD A TOE SHOT AWAY.
AND WAS
GASSED AT NEUVE CHAPELLE
AND INVALIDED HOME.
PLEASE GIVE ME PENNIES TO SEND
OUT CIGARETTES   *Signed* JACK.

FIGURE 21. Monkey mascot Jack at Bristol Zoo, c. 1916. Viner
& Co. Ltd. (Roy Vaughan Collection, Bristol Zoological Society
Archive, courtesy of the Bristol Zoological Society)

mate and human worlds. In a 1928 description of orangutans, it was noted
that "the female is smaller and more active and is always at the beck and call
of her lord and master, for whom she often fetches food and drink." Similarly,
a female chimpanzee who had recently given birth to a young male called
Adam was described as "triumphant . . . almost a smile of satisfaction [lies on

her] kindly features." Later she was styled as a perfect mother, guarding Adam with her life.[49] Similarly, the family as a typical social unit in which the male took ownership of the group was often imposed. In 1945, Jack, a chimpanzee, was described as proud of his family as a "model of domestic harmony." Indeed, males were generally represented as detached from family life, concerned chiefly with sex and recreation, playing no significant part in rearing their young, though depictions of loving heterosexual relationships were not uncommon, either.[50] The identification of common ground in these ways actively encouraged an understanding of animal life which broke down the nature/culture binary and which promoted affectionate/empathetic looking.

Slightly mawkish newspaper reports mourning the loss of animals, usually mammals, were also increasingly deployed as ways of recognizing and promoting animal individuality. The death of the Zoo's charismatic megafauna was big news. Two polar bear cubs, sent for treatment at the Bristol Royal Infirmary in 1927, failed to recover. One newspaper lamented that their passing would inevitably be a source of enormous sorrow, not only for their parents "Mr. and Mrs. P. Bear," but also for visitors to the Zoo. The death of a chimpanzee called John in 1938 likewise induced deep mourning both from visitors to the Zoo and his "disconsolate widow," Betty.[51] Many of these kinds of press reports entrenched this appreciation of the loss of a *person* by relating brief digests of their captive lives (though by no means to the same degree as might be found in human obituaries). The popular chimpanzee Markeene, for instance, was the subject of a newspaper obituary in 1935, which related her taste for cigarettes and her past roles in films produced at Britain's famous Elstree Studios.[52]

This kind of anthropomorphic representation and emotional perception of animal personality became commonplace over the course of the period between 1900 and the 1960s. The cultural domestication of animals in zoos reached its zenith, however, in the deliberate propagation of celebrity beasts. These animals' assembled personalities were developed to such a degree that they became more than merely familiar strangers. Instead, they were transformed into honorary people who were tightly enmeshed within the Zoo's and indeed the City of Bristol's social fabric. The rise of a British culture of celebrity can be traced back to the late 1840s. Indeed, the rise of the phenomenon of celebrity predates the late nineteenth or early twentieth century's expansion of media technologies and increasing levels of literacy among the general populace. These factors, however, expedited a gathering curiosity in, and enhanced access to, the intimacies of the lives of particular characters, animals included.[53] Increasingly, from around this period, zoo animals were not merely famous as transitory wonders or especially popular exhibits. Rather, they were frequently subject to a (variously) prolonged public interest in which they came to be known, and in some cases loved, as somebodies. Most notably, the hippopotamus Obaysch (1850–78), and the elephant Jumbo (about 1860–85) were both displayed at London Zoo and were equally subjects of an intense public fascination through which they were symbolically domesticated. Both were awarded distinct almost-human personalities. These identities, however, were distinct in their coloration in comparison to the animal celebrities of the twentieth century.[54]

While Bristol Zoo displayed a number of famous lions in the 1800s, individuals known as Jupiter and Hannibal among them, it could not boast national pets of the same hue as Jumbo or Obaysch until well into the twentieth century when the Society was intensively propagating such animal identities. A number of animals emerged at Bristol in the 1930s, such as Alfred the gorilla, Adam the chimpanzee, and Rosie the elephant, whose lives illustrate the ways in which some animals, always charismatic megafauna, were subject to an intense form of cultural domestication and emotional looking.

Rarity aside, an anthropocentric animal biography, located in the exotic wilds of Africa but which marked Alfred out as a special individual, deepened the exhilarating dimensions of his physical form when he arrived at the Zoo in 1930. Alfred had supposedly been breast-fed by an indigenous woman after he had been orphaned in West Africa. American adventurers W. K. Gregory and Henry C. Raven later happened upon him in the possession of a Greek merchant in Mbalmayo, Cameroon, in 1929. They claimed that they witnessed him drinking milk from a can and were most amused by his various antics.[55] Alfred spent eighteen years at the Zoo during which time he was routinely constructed and recognized as a somebody with a very distinct personality. Because he was about two years of age when he arrived, the Society could effectively present him as an endearing child-like character. He was often dressed in human attire, so a manufactured human persona was laid on top of his status as a rare representative of the species *Gorilla gorilla gorilla* (fig. 22).

FIGURE 22. Alfred the gorilla with keeper, 1930. (Roy Vaughan Collection, Bristol Zoological Society Archive, courtesy of the Bristol Zoological Society)

Alfred lived his life in the public gaze. Indeed, Bristol's newspapers regularly conveyed his story to their readers. Portrayed as most appreciative of the attention lavished upon him, he assumed "the air of monarch of all he surveys." In 1934 it was reported that "up and down his quarters strutted Alfred, a very king of gorillas, thumping his chest and making strange sounds not unlike 'look what a fine boy am I.'"[56] Of course, Alfred was on display at the same time as a giant gorilla terrorized New York on film in 1933's *King Kong*. This might explain the "king" reference (though, of course, it could also refer to King Alfred of burned cakes fame). This served the purpose of connecting the two creatures within cultural consciousness.[57] The Society also publically celebrated his birthdays with cake, and in 1935, he had his first "shave" (which referred to him starting to pluck his eyebrows and the hair on his chin: this is a stereotypical behavior common among gorillas and other primates in captivity, and it is an activity that we now understand to be a symptom of psychological disturbance).[58] Such was the media coverage surrounding Alfred that in 1937 the Society was able to claim that he was the most famous gorilla in Europe.[59] The longer he lasted at the Zoo, the more exceptional he became. His longevity, like Bushman's in Chicago (1930–51), significantly enhanced his status, especially in light of William Temple Hornaday's bold declaration in 1915 that it would never be possible to maintain a gorilla for any length of time in captivity.[60] More than that, however, his longevity allowed his fame to spread and deepen.

The outbreak of the Second World War was integral to the heightening of Alfred's reputation, not only within the city of Bristol but also far beyond. As one of the few large animals to remain following the depletion of the collection through euthanasia, sale, and donation, he not only became one of the few charismatic figures onto whom the visitor's gaze could be cast but also a figure who, like Britons more broadly, would never surrender. His popularity in Bristol and the challenge posed by his considerable appetite in light of wartime trade embargoes heightened his fame, which became global by 1940. His plight received coverage in newspapers in North America, Australia, and New Zealand, while US GIs stationed in Bristol were specifically encouraged to take a visit to the Zoo to see him. These men supposedly sent thousands of postcards bearing his image to their families back home, extending his fame across the pond.[61] Referring to the rationing of Alfred's food, and employing the Hollywood terminology of "Glamour boy," a 1946 poem expressed a sense of his being under the intense glare of the spotlights:

One isn't generally nude
When one is being interviewed
But ALFRED in his birthday suit
Spoke longingly of milk and fruit
Of grapes and oranges galore
He used to get before the war;
But now the only fruit he had
Was generally going bad
Which rather whittled down the joy
Of being BRISTOL'S GLAMOUR BOY[62]

His popularity and alleged hunger during wartime meant that, like the First World War monkey mascot Jack, for some, Alfred became a symbol of life endured in the shadow of war. He was "one of the boys," struggling valiantly against the evils of Nazism. His nickname—"the Dictator of Bristol"—elevated him as an emblem of resistance in the face of the Fascist threat gaining ground in Continental Europe. His construction in this way was particularly potent in light of the bombardment Bristol endured during the Blitz in 1941–42.[63]

In a variety of ways and for a variety of reasons rooted in significant social contexts, then, Alfred was elevated above other captive creatures as an almost-human companion creature of civic, national, and increasingly global importance. Over the course of his life, he became a locus of emotional attachment, particularly among children. The construction of personhood in juvenile relations with animals is commonplace in our culture, and such a conceptualization represented a significant foundation in the creation of Alfred as a celebrity beast. He was regularly construed, and is still remembered, as a much-loved personal friend, who was capable of looking back, making connection, somehow comprehending the human condition from across the chasm of the human-animal divide.[64] His death in 1948 was a moment of great distress; such was the connection many people, especially the young, felt toward Alfred as a some*body,* rather than a some*thing.* Of all the Zoo's animals, Alfred received almost as much media attention in death as he had in life. His demise from tuberculosis was reported extensively in the Bristol press, his final hours recounted as they might have been in a human obituary. His last meal was apparently an orange, followed by a well-deserved bottle of stout.[65] Following his death, his body was reconstituted for display in the Bristol Museum, and a bust of his head was carved in stone, his presence at the Zoo forever enshrined in a kind of grave etched into the zooscape.[66]

The captive life of the chimpanzee Adam similarly reflects—and further nuances our understanding of—the ways in which zoo celebrity and personhood were fostered. When he arrived in 1934, he was the first chimpanzee ever born in captivity in the UK. Like Alfred, he was possessed of a quality of novelty, and also like Alfred, he entered the public imagination as a childlike figure. An almost-human life was depicted through events to which children might relate. Iced birthday cakes were prepared for parties to which children and human celebrities were invited. In 1935, he shared his birthday with "Peter Pan," otherwise known as actor Jean Forbes-Robertson. Hearing J. M. Barrie's famous children's tale and the exhilarating adventures of the Lost Boys in Neverland, the young chimpanzee apparently became frightened of Captain Hook; he entreated Forbes-Robertson to reveal to him the secret of flight so that he might escape. On his second birthday he was "thoroughly naughty," like a "little boy," though he still received his birthday cake alongside postcards from children across southern Britain.[67] Adam was well on the way to cementing his own celebrity when he died in 1940 at the age of six. Because of his early demise, he did not rise to the status of Alfred. In other words, he did not quite threaten to traverse the chasm of the human-animal boundary.

The primate-ness of these animals was important, and Alfred, at least, really was exceptional in the magnitude of his celebrity at Bristol. Even other popular mammals were not perceived as almost-people to quite the same degree. The elephant Rosie, for instance, arrived in 1937 and was a renowned and well-traveled elephant of the stage, having featured in the 1937 film *The Elephant Boy* and having performed in productions at the London Coliseum and the London Palladium. As a riding elephant, she was unendingly popular with children. She was described as the "children's own," just as London Zoo's elephant, Jumbo, had, in a previous century, been dubbed the "children's friend," or "children's giant pet."[68] Like Alfred, Rosie endured the war in Bristol, but she was neither a young animal nor primate. While she was certainly a zoo pet, she did not muddy the waters of the human-animal borderlands as a great ape such as Alfred could, and so she was never subject to quite the level of affection afforded Bristol's very own "Dictator."

Thus, the intensity of visitor affection toward Alfred and, to a lesser degree Adam was a result of the fact that they were not merely mammalian but also primate. Such creatures were consistently popular with zoo-goers. Out of 203 animal adoptions between 1940 and 1941 (a means of raising money during wartime), 119 were of mammals, and, of these, 45 were of primates.[69] By the end of the nineteenth century, attempts to discern fundamental distinction between human and anthropoid ape had become less marked in the Society's official literature, commensurate with a changing worldview focused less on the privileged place of the human among the marvels of Creation and more on the processes of evolution and the realities of species proximity in an interconnected and interrelated biosphere.[70] Indeed, the Society went to great lengths to maximize the humanity of its primates during the twentieth century and particularly from the 1930s. Great apes were routinely presented to the public through the ascription of human forenames such as Betty, John, Timothy, Helen, Joan, and Paddy. These names alone promoted and entrenched the animals' individuality and potential proximity to the human condition.[71] For a time, too, this depiction of proximity was further enhanced through the clothing of the animals in human apparel or the mimicry of human activities like smoking cigarettes (fig. 23).

Amid the evocation of a familiar kind of individuality, there often lurked

FIGURE 23. Lady Clara with keeper, c. 1931. (Roy Vaughan
Collection, Bristol Zoological Society Archive, courtesy of the
Bristol Zoological Society)

a subtext of inferiority in the representation of primates, which substantially diminished their prospective personhood. In the process, their status as
liminal creatures of an in-between realm remained beyond doubt. Tea party
performances were one way in which their likeness to humans was addressed.
Simultaneously, through the disorder that usually ensued, such spectacles

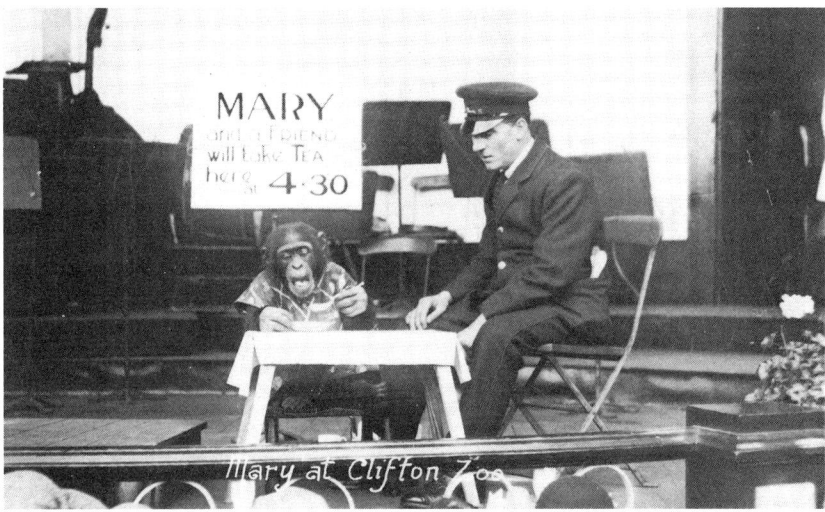

FIGURE 24. Mary with Ralph Guise at Clifton Zoo, c. 1927. (Roy Vaughan Collection, Bristol Zoological Society Archive, courtesy of the Bristol Zoological Society)

portrayed them as emphatically *not us*. During the summer months of 1927 a well-known chimpanzee named Mary visited the Zoo. Her owner, writer and early wildlife photographer Cherry Kearton, treated her like a human child. During performances, which consistently attracted a huge crowd, she would drink tea and eat toast, cake, and bananas, before grooming herself and then larking about on the zoo lake (fig. 24).[72]

Great apes, along with primates more generally, were able to evoke among visitors a potent but unstable sense of species proximity, and this recognition was often joined by unease. A description on the back of a postcard featuring the chimpanzee Lady Clara, wearing a hat and smoking a cigarette, noted that "this monkey was so funny . . . it did exercises for its food and held a mug out to be filled, and drank like a man."[73] Comically demeaning in nature, these remarks disclose an ambiguous and uncertain perception of these boundary-straddling creatures that so easily prompted feelings of empathy, affection, and confusion in equal measure.

The cultural domestication of animals as recognizable individuals intensified even further in the years during and following the Second World War. Increasingly, this was achieved through explicit contextualization within a children's world. More than ever before, animals were positioned by both the

Society and the mass media as friends who consciously engaged with children. While animal mascots were previously attached to causes relating to international conflict, they now became involved with issues of particular concern to the city's children. The chimpanzee Timothy became the mascot of the *Bristol Evening Post's* Pillar Box Club for Children in the late 1950s. His image was deployed in order to raise funds for local youngsters who had physical disabilities. Similarly, the young gorilla Daniel became the mascot of the *New Observer's* Children's Club in the early 1970s.[74]

The growth of television presented new channels through which the Society could promote this kind of image of its animals. The kind of mass communication afforded by the television screen allowed images of friendly animals to be beamed right into the living rooms of people all over the country. First broadcast in April 1962 and hosted by radio and television broadcaster Johnny Morris, the BBC children's serial *Animal Magic* ran for twenty-one years and was frequently filmed at Bristol Zoo. The program depicted individual animals and, by extension, whole species as friends of human children, leading inner lives that were not foreign to the lives of youngsters slumped in armchairs and on sofas across the nation's living rooms. In an episode broadcast in December 1962, Morris was filmed hanging stockings for monkeys, while describing the elephants, Wendy and Christina, as "naughty" children excitedly awaiting the arrival of Christmas. Jovial festive music accompanied the visuals, presenting the animals, and their relationships with Morris, as harmless, amiable, and amusing. The creatures with which Morris shared the screen were unambiguously cast as almost-human pets; tigers were often referred to as "pussy cats," young animals as children, and human voices and accents were assigned to them so that they seemed to talk directly to both Morris and their juvenile viewers.[75]

As we have seen, however, these kinds of animal persons were usually mammalian. Even in the midst of rising emotional and anthropomorphic looking, animals lower down on the hierarchy of animal life usually rigidly remained objects rather than subjects. Early twentieth-century photography illustrates the ways in which physicality rather than emotional worlds were emphasized in relation to such creatures (fig. 25).

Well beyond the middle point of the twentieth century, when mammals had become individuals of intense sentiment in zoos, fish remained somewhat divorced from the rest of the animal collection as objects of beauty. In

Boa Constrictor.
Born in the
Clifton Zoo. 1901.
1453.

FIGURE 25. Boa constrictor, 1901. Viner and Co. Ltd. (Roy
Vaughan Collection, Bristol Zoological Society Archive, courtesy
of the Bristol Zoological Society)

so doing, they were demarcated as something "Other" entirely. As with the original Aquavivarium, which had been established around 1856, when the Society opened a new aquarium in 1927 they levied an initial surcharge of six pence for adults and three pence for children. This additional charge remained in place until the end of 1961 and served two purposes: it contributed to the significant capital expenditure of the exhibit and symbolically separated the creatures therein from the rest of the animal collection, positioning them almost as odd breeds of animate ornament.[76] At the opening of the aquarium, reports emphasized the "splendid," "vivid," "beautiful" collection of fish. One newspaper enthused that "the new aquarium at the Bristol Zoo is a source of never-ending delight . . . some have marvellous fins like butterflies' wings; some enormous heads like bufalloes . . . they are really rather beautiful and very graceful in their movements." Indeed, the decorative conceptualization of fish is consistently apparent in official twentieth-century publications. Guidebooks from the late 1920s through to the mid-1970s predominantly present them in terms of their appearance and functionality, from the teeth of the piranha to the spectacular forms of the angelfish and the colors of the goldfish. There is neither consideration of individuality nor any trace of feelings toward them.[77]

In similar fashion, and following the declining perception of antagonism between the human and animal worlds, twentieth-century depictions of reptiles focused on the beautiful forms and structures of the creatures, fixing the gaze on aspects of otherness rather than any kind of similarity. Yet, reptiles persisted as a complex case, for in many ways they remained "bad" animals. They were objects of negative portrayal in the press and objects of terror and fascination in the public imagination.[78] A 1936 newspaper report labeled a group of juvenile snakes "very spiteful," and as late as 1963, the *Bristol Evening Post* related the crocodile's reptilian cunning and proclaimed that his eye "looks about as baleful as anything in nature."[79] In a poem written by visitor Peter Nichols, in which he reflects on boyhood visits to the Zoo during the 1930s, he recalled the "foreboding doors" of the Reptile House, beyond which "there lurked a sense of evil" and "creatures of my darkest dreams."[80] Indeed, a 1963 newspaper reported that, despite a general attitude of "hatred" toward the creatures, the snake house was one of the most consistently congested parts of the Zoo due to the public fascination in the ability of the reptiles to consume their prey whole, as well as their enjoyment of the creatures' strange forms and behaviors.[81]

While connection to a human world was a vital part of the Society's presentation of some of the animals inhabiting its Zoo, genuine relationships founded on the recognition of animal subjectivity continued to flourish between keepers and kept. Material and emotional intimacies developed between humans and individual animals across the period, though the interface between senior keeper Mike Colbourne and the great apes he managed is perhaps the most powerful example of these kinds of attachments. Because he was "more of an ape person than a human person," his intense and immediate maintenance of great apes allowed him, he claimed, to learn to speak gorilla. Frequently entering the enclosures to play with the animals and "earn their trust," he tried to integrate himself into primate groups. He was an "almost constant companion" to the gorilla Delilah after she had given birth to Daniel in 1971. He also became something of a surrogate parent to young great apes, particularly the orangutan Oscar (1971), gorilla Daniel (1971), and the gorilla Naomi (1977).[82] For Colbourne, each of these animals was an individual he might know and love. The degree of contact Colbourne was allowed with the great apes is astonishing by today's standards. He would lie on the floor of the enclosure stroking and speaking to the pregnant gorilla Diana as she laid her head on his lap. Speaking of this relationship, Colbourne referred to a mutual confidence, recalling that he was able to control her anxiety by squeezing her hand; such was the degree to which they knew each other.[83] Through the variety of nurturing roles Colbourne performed in interactions with great apes, he embodied relationships with his animal wards that moved beyond representative scientific or sweeping anthropomorphic conceptualizations. These animals were individuals with whom Colbourne believed he intimately related and whom he loved deeply. The recognition of animals as individuals, or even as friends, then, was not something that was imaginary. On the contrary, intimacy and proximity often featured prominently in the day-to-day care of captive animals, even at the height of the twentieth century's anthropomorphic partialities.

PETS FOR PRESERVATION

As we know, the final decades of the twentieth century witnessed seismic changes in the character of zoos. The intensification and expansion of a conservation ethos influenced the ways in which the Society related to its exotic captives. An integral part of this reorientation involved some degree of rejection of the extreme forms of individualization and anthropomorphic looking that had coursed through relationships with animals during much of the previous decades. Such attitudes and behaviors toward animals were increasingly considered distasteful for they risked disconnecting zoo visitors from the realities of animal life and the very physical threats facing them in the wild places. Despite this, however, recognition of individuality and certain kinds of emotionally driven looking endured at the heart of late twentieth- and early twenty-first-century relationships with wildlife. While captive animals of this period might have been less explicitly "pet," many of them nonetheless endured as creatures straddling the human-animal borderlands, members of an eclectic multispecies family at home at the Zoo.

During the final decades of the twentieth century, extreme cultural domestication through anthropomorphism became generally less acceptable within both entertainment and educational contexts.[84] Indeed, the BBC's ultra-popular *Animal Magic* had fallen out of favor as a mode of presenting animal life to young audiences by 1983.[85] And yet the fostering of visitors' emotional connection to captive animals remained vital not only to the Society's commercial imperatives, but also to the communication of its emergent conservation message. If the Society was to effectively challenge and then transform everyday human behaviors, it needed to encourage visitors to the Zoo to care about the animals they encountered. Much of this has been achieved through the construction of individual animal ambassadors who reflect the plight of their species and who remain recognizable as objects of emotional knowledge.

One method of achieving this has been through the continued gendering of animal life. Though this is less explicit than it once was, the recognizably maternal love of the gorilla mothers, for instance, is still emphasized, while the Society's present silverback, Jock, who was born in 1983 and acquired from Zoo de la Palmyre, France, is cast as "the protective dad," thus perpetuating the division of gendered roles according to familiar social norms.[86] Naming practices have also endured, but while the ascription of individual names continues to denote the presence of an individual animal with a rich life of its own, the names assigned speak less of human-pet relationships, less of the inferior, passive, cuddly animal other, and more clearly of the careful creation of a lively conservation ambassador who can arouse an emotional gaze and embody the physical threats facing the wild world. One of the most recent primate arrivals at the Zoo, a gorilla born in September 2011, was named Kukeña, meaning "to love" in Congolese. This is quite distinct from most former names afforded to great apes such as Jack, Jill, Alfred, or Henry.[87]

The ways in which the Society has dealt with animal mortality in recent decades similarly reflects this shift in relationship. Death not only marks the departure of an individual but has also come to be emblematic of the loss of species amid the horrors committed at the hands of the rapacious human animal. Press releases relating to the losses of Asiatic lions Moti in 2010 and Kamal in 2012 illustrate both the endurance of a presentation of animals as knowable individuals alongside an emphasis on the impact of the losses on the plight of a critically endangered species.[88]

Alongside this, affection continued to result from the recognition of animal individuality among workers at the Zoo. This is in spite of a departure from earlier practices whereby keepers were increasingly prohibited from engaging in hands-on interactions with their animals in line with the conservationist paradigm underscoring the Society's work to preserve precious animal bodies. It became important for future reintroduction programs that captive creatures remain as "wild" as possible, even in their artificial worlds. Keeper Neil Clennell remarked in 1994 that he had become "surprisingly attached" to a five-hundred pound tortoise known as Big Tort, while keeper of mammals Melanie Bacon (née Gage) noted with reference to a young male gorilla, Djengi, that "it's been like having a baby. He has boundless energy and an infectious giggle."[89] Staff members apart from keepers also developed emotional bonds to animals they encountered in the course of their daily activities. One recalled a number of "magic moments" alone with a polar bear called Misha. He would, she said, now and again look up at her, returning her gaze from the depths of his pit. Another recalled spending time alone with the elephant Wendy, who would often salute her by raising her leg and trunk.[90] In these various ways, and from various vantage points, zoo staff continued to see captive animals as recognizable individuals who were unique within the body of their species and who could be known on an emotional level.

Visitors, too, continued to develop intense bonds with certain animals. The rise of animal rights movements in the final quarter of the twentieth century denotes an increasing disruption of the boundary separating feeling, thinking humans from "dumb brutes." Suffering became an increasingly potent consideration in the imagination of animals as sentient beings. In 1974, for instance, the *Western Daily Press* reported that the popular lions Bess and

Dougal, a tiger, and a puma were to be euthanized. The tone of the report was anguished; the animals were "doomed" to having their lives unjustly brought to an end, it lamented. Intrinsic to the protests of a local animal rights group, Bristol Zoo Action, around 1993 were objections to the deaths of zoo animals that, they alleged, had come about from their ingestion of infected food, their being gored by other animals, and their consumption of gravel in their enclosures. All of these caused the animals to experience extreme suffering.[91]

Consequently, sentimental feelings toward animals on the part of the public and zoo staff have in some ways intensified at the same time as the Society has adopted less overt anthropomorphic representational devices. The 2002 euthanasia of the elephant Wendy after forty-one years at the Zoo ended a long tradition of elephant display in the Gardens and nicely exemplifies the way in which sentimentalization and objectification sat side by side in human relationships with captive animals. To all intents and purposes, Wendy's death revealed her to be a *person* to whom staff and visitors felt a powerful attachment. The manner of her demise itself was about the manifestations of compassion in the midst of gruesome realities. The decision to euthanize, in light of her extreme and prolonged suffering from a painful degenerative joint condition, was reported as the "only responsible" course of action. The slaughter itself was an occasion that the Society managed meticulously. At six o'clock on the evening of 11 September, a select group of staff met inside the elephant house, where a five-milliliter M99 etorphine hydrochloride ("immobilon") dart anesthetized the creature. One liter of potassium chloride was then injected into a prominent vein in her ear, easing her passage from life to death. It was anticipated that Wendy's passing was to be quiet and respectful, reflecting the emotional tumult many staff felt in slaughtering the animal even if it was so that her suffering would come to an end. Once her death was confirmed by electrocardiogram, additional staff and members of the veterinary team were permitted to enter the enclosure, while distraught colleagues who had been present at the moment of her departure were allowed to leave. A ring of shavings was laid down around the carcass to collect fluid, and equipment was assembled in preparation for the night's postmortem activities. Gloves, wheelbarrows, large knives, a grinding stone (a member of staff was put on knife-sharpening duties since elephant hide quickly caused blunting), an electric saw, brooms, shovels, and scalpels were prepared and the drains sealed to prevent pachyderm blood entering subterranean watercourses. Refreshments were provided under the Pelican restaurant canopy. One former maintenance worker noted that it was "quite a sad thing when we were cutting her up." At six o'clock the following morning, the skeleton was loaded onto a truck and transported to Edinburgh to be stored as part of a vast collection of research specimens, while the fleshy waste products were removed from the site in two hundred 8-kilogram bags weighing a total of 1.6 tons.[92]

The public reaction to Wendy's death was intense. In an outburst of emotion that harked back to the torrent of public grief that accompanied the death of Diana, Princess of Wales, only five years earlier, memory books were opened in which people could reminisce or wish Wendy the best for her afterlife.[93] Six books were completed featuring more than 3,500 messages from staff and the public, and they reveal an array of emotional relationships with the animal in the contexts of both life and death. For some she was a friend, to others a victim of abuse through captivity. Most reinforced the sentiment that she was loved and would be missed and remembered forever. "When sweet winds blow," one person promised, "I will always hear the name Wendy. God bless you." For many others the conceptual boundary between human and nonhuman dissolved in their belief that she possessed an immortal soul, which had now moved from the confines of her fleshy body into the hauntological domain of an infinite afterlife. "I think Wendy has gone to a sunny place," one remarked. Some recalled memories of the elephant in and around the Gardens, while others professed their relief that she was, in death, spared the horrors of a captive life.[94]

Once she had been mourned, her individual life was memorialized. Her image was carved into the trunk of a Lebanon cedar tree in the Gardens. Her wooden body provided the base for carvings of a penguin, Galapagos tortoise, Asiatic lion, and pygmy hippo. All of these animals are endangered species, and so Wendy's life was physically remembered in the compelling context of the Society's contemporary mission to rescue and defend the wild things and wild places of the world. In an age of social networking, Wendy's memory also came to lurk among the pages of Facebook. Visitors to the group were asked to "leave Wendy messages . . . remember the good times . . . and generally pay respect to everyone's favourite elephant." Over three thousand messages have been posted on the page to date, most of which echo the sentiments of the memory books opened in the days immediately following her demise. However, the web page also reveals attitudes toward the animal that were derisory and sometimes vulgar, illuminating a multiplicity of attitudes toward the dead creature in afterlife: "Wendy sucks" wrote Trevor Campbell; Amelia Fuell reported that "Sam Mathews used to spend a lot of time with Wendy, if you know what I mean ☺"; Mark Holloway recalled that "I remember taking Wendy out on a date, we headed for a little restaurant tucked in the corner of town, she looked Stunning. After our meal we decided to head to the cinema, to see her favourite movie, Dumbo."[95] Some of these messages may be disrespectful, but they nonetheless reflect the enduring evocation of the memory of an individual creature.

Again, however, this kind of human-animal relationship remains confined, broadly speaking, to mammals, and this reflected the endurance of the hierarchy of animal life, which so predominated across the nineteenth and twentieth centuries. The Society expressed despondency at the passing of the lioness Mary in 1994, devastation at the death of the female western lowland gorilla Undi in 2000, and "great sadness" regarding Wendy's death in 2002. The demise of a dwarf crocodile in 2001, in contrast, was merely "unfortunate," the sense of disappointment assuaged by the fact that replacements would soon be acquired.[96] As in previous years primates retain a particularly strong hold on the imaginations of the Society and visitors to its zoo. From the final decades of the century, apes were portrayed less as semi-comic almost-people and more as humans' closest genetic relatives, reflecting the changing significance of species proximity in scientific studies and philosophical and ethical debates.[97]

In this context, charismatic megafauna have a critical role in the transmission and delivery of the aims of the conservation agenda. The Society's present family of western lowland gorillas are each known by name, and each has their own family roles, perpetuating dissolution of the borderlands separating humans from the rest of the animal world and encouraging emotional responses to the plight of the species both in captivity and in the wilds of Western and Central Africa. And, indeed, they are deployed as mascots, emblems of extinction—compelling action for all the endangered species of the world. Evoking the memory of Alfred in some part, the Society's superhero, Kuki (after the young gorilla, Kukeña, born in 2011), is deployed as a means of focusing attention on the plight of the Garden's threatened species. "Quick, help!," he shouts in one instance, "the little *patula faba* snails are in danger.

They are being eaten by snail-eating snails that don't belong there . . . extinction is coming for them on their South Pacific island. Let's save them by flying them to the Zoo for safety." In so doing, the image of the gorilla is used to encourage zoo-goers to care, not only for popular captive creatures, but also for "lesser" zoo animals and their wild kin. These flagship species transfer affection from their own bodies onto the forms of other species, and, thus, the Society expands the moral circle entrusting zoo visitors with the responsibility of caring for a fragile world.[98]

This is important because there are plenty of species, snails among them, that still possess limited emotional leverage. Indeed, they are less easily recognized as individuals among the mass of their species. Fish still remain objects of beauty rather than individual subjects of vibrant lives (just look at pescetarianism as an alternative to a vegetarian lifestyle). In 1990 a young girl wrote to the Society, marveling at the colors of the fish she saw on her visit, while a few years later another visitor wrote to say that "I'm so glad to see the sea horses in the Aquarium—it is very fascinating to watch them move. They look like mythical creatures, don't they?!"[99] Reptiles similarly remain objects of fascination and fear in the public imagination, rather than as recognizable subjects of rich individual lives. Indeed, the Society itself acknowledges the perpetuation of such widespread negativity, having begun to deliver snake phobia (alongside arachnophobia) courses in 2009.[100] These kinds of creatures do not generally evoke positive and empathetic emotional responses, which is problematic when species' survival it at stake.

CONCLUSION

The ways in which people imagine captive animals—the Society, the press, and zoo visitors—reinforces the sense that shape-shifting beasts have always populated the Zoo. On top of the layers of relationship that objectified animals were also relationships that recognized animals as subjects, and this made them the foci of an array of feelings. Such sentiments denote the recognition of the presence of a living being with a life of its own. In many ways, these creatures were figured into people, if you will, who each deserved recognition within a moral circle that was wide enough to embrace species other than *Homo sapien.* This way of relating to animals forms a layer of relationship that, sometimes perplexingly, lies alongside cold objectification and the exploitation that entails.

With animals often interpreted as humans in animal skins, these ways of looking at animal life were decidedly fluid and heavily contextual, changing over time according to species and often reflective of individual experience. A sense of antagonism and human exceptionalism grounded these relationships with animals during the nineteenth century, though these ways of understanding the human place in the universe dissolved in some degree once Darwin's insights and the rise of mass media allowed for an increasingly positive consideration of some animals as people like us. The image of the animal as recognizable friend was a vital part of the Society's self-image and allied appeal over much of the twentieth century, and it was milked for all it was worth. Later, the rise of conservation and ecology refigured those relationships. Yet, the recognition of individuality and the emotional and sometimes anthropomorphic looking that resulted nonetheless remained at the heart of the human-animal relationship at the Zoo. Empathy for animal life was the key to promoting their care in a world on the brink of disaster.

Despite the recognition of individuality, the capture, management, display, and imagination of animal life are all forms of appropriation, interpreting animals through frames that are almost always informed by our own practical and emotional assumptions and desires. To imagine that such appropriation equated to total and uncontested human domination of animal life at the Zoo would simply be to appease the metanarratives of domination, oppression, and exploitation promoted by those opposed to zoos in principle, however. In fact, a significant feature of human-animal relationship at the Zoo over time has been the ability of animals to shape the zooscape and their interactions with human others within and beyond it. Captive animals were frequently active parts of a profoundly multispecies space, as well as objects of intense imaginative manipulation. It is to this aspect of the status of captive animals to which we now turn.

5

The Power of Nature,
and the Natures of Power

As the long, hectic summer season of 1944 drew to a close, the popular riding elephant Rosie started behaving erratically as she walked along the Zoo's Main Terrace. Without children on her back but with visitors all around, low-flying aircraft had spooked her. She reared up on her hind legs, momentarily evading the control of her long-standing keeper, Tom Bartlett. Though no injuries were inflicted on this occasion, Secretary Dr. Richard Clarke was acutely aware of the threat the elephant posed to human life and limb if she were to entirely break free of the mastery of her keeper. He duly contacted the Bristol Aeroplane Company, warning them of the consequences of low flight paths in airspace over the Zoo. This was not the first time an animal had reacted badly to the aerial incursion of modernity; the celebrity gorilla Alfred had supposedly disliked airplanes flying over *his* zoo in the late 1930s and early 1940s. Three years later, in 1947, Rosie collapsed in the intense heat of another long summer season. This time she was carrying a group of children on her back. While they suffered only minor cuts and bruises, the elephant was consigned to her enclosure to recover, her steadfast keeper Bartlett remaining at her side. Indeed, he firmly believed that she had intentionally saved the lives of the children by lowering her body gently to the ground rather than allowing herself to fall. A further two years later, Rosie flicked her trunk between the bars of her enclosure, striking and concussing a seventeen-year-old boy. Another moment of pachyderm action certainly, but this one was different. Rosie had compromised the safety of the Zoo's young visitor by initiating violent contact across the threshold separating boy and beast.[1] These episodes and the discourse surrounding them certainly reinforce the idea that there are many layers to the human-animal relationship. More than that, however, these three events, both separately and altogether, also show that Rosie had the capacity to be a powerful influence on people and place. This power, however, was itself subject to manipulation: she was, for instance, not necessarily seen as culpable for the accidents outlined above. She had lived at the Zoo between 1937 and 1961, at a time, as we have seen, of intensifying sentimental and anthropomorphic recognition and depiction of captive beasts. At this moment, they were almost unambiguously cast as friends compliant in the delivery of wholesome amusement. Retributive action against this elephant was unthinkable.

Human lives are bound to such moments of animal action and effect. Zoo animals have been, and still are, generally conceived of as passive objects of exploitation, helpless in their captive states. This is hardly surprising, for the layers of relationship we have already uncovered reveal a menagerie that was remarkably malleable and subject to an array of unstable human hallucinations. Captive beasts were bent according to human whim: precious beasts were acquired and positioned in cages from which they were not supposed to escape and in which they were explicitly objects to be relentlessly observed within an anthropocentric frame of reference. They were seen too as multisensory entertainments, feeding, riding, and petting being manifestations of mastery, for they rested on the ability of one species—humans—to initiate and manipulate bodily interactions with animal others. Beyond that, their bodies and behaviors were excavated and secrets extracted while, at the same time, they were thought of as subjects-of-a-life often reflecting back at us our own brittle sense of self. All of this took place at once and in intricate, unstable and transforming ways over the period. Such interspecific relationships in this kind of artificial environment seem hardly to be illustrative of a negotiable balance of power.

These physical and figurative relationships, and the recontextualization of captive animals inherent within them seemingly render the beasts passive and at the mercy of the endlessly manipulative human animal. As the three episodes relating to the elephant Rosie suggest, however, such an imagination of life in captivity fails to tell the whole story. Living things always interact with the world around them. They are part of what Tim Ingold has called, in his critique of Latour's conceptualization of "networks," a "meshwork": a complex—and concrete—world of entanglement.[2] To imagine a creature made passive is to conceive of a fragment of a mosaic in complete isolation, without any relationship with, nor capacity to influence, its environment in however seemingly insignificant ways. Looking back at the humans, animals, spaces, and places that we have encountered in this book, we might think of the zoo, then, not only as a space in which human fantasies about animals and the rest of the natural world manifest, but also as a hybrid space, an eco-system in its own right. It has always been, from its very inception, a place of multispecies making, and thus of a shifting and unstable balance of power. Whether or not that balance was equal, of course, is another question entirely, and, indeed, that is not the point that I am trying to make. This is important. David Gary Shaw recently posited that "possessing agency brings to the fore the very purpose and point of history. For some, to deny historical agency seems almost to deny historical significance. Indeed, one reason for expanding history's subjects to include the whole of human society was to demonstrate that agency could exist where political power was insignificant."[3] By identify-ing fragments of nonhuman influence, we can better understand the past as something that is complex and multiauthored. Moreover, we can conceive of historical human-animal relationships as, in some respects, cocreated.[4]

The problem here is that the term "agency" is loaded. It is inextricably as-sociated with rationality and the intent to achieve predefined goals. Applying the concept to animals is therefore problematic; instead, the term "influence" better serves the purpose of illustrating the ways in which animals can copro-duce the world, whatever their motives may or may not have been. Thus, this chapter shows that captive animals across the period were influential beings involved in an interactive process of making space and forging relationships. They affected bodies, crafted attitudes, and helped create intimate interspe-cies interactions. Often, they did not do as they were told. They escaped, bit, spat, and refused to reproduce or to eat their prescribed foodstuffs. In so doing, they impacted on the multispecies life of the Zoo. Yet the Zoo was a space in which power circulated and was negotiated. It fluctuated, zipped back and forth among living beings, both human and more-than-human. In-deed, when an animal exerted its influence, it was rarely left to stand for itself.

Animal behaviors were interpreted and reinterpreted in order to explain unusual or undesirable behavior; and physical measures were put in place in an effort to wrestle back control of animal actions. Creaturely influence was frequently subsumed into an anthropocentric system of meaning making, where animal influence was shaped to serve human ends. What we find when we look at the Zoo as a vibrant multispecies environment, then, is a world in flux: living things, human and more-than-human, perpetually impacting on and reframing the relationships between them.

That being said, the complexions of these interactions were nonetheless inconsistent across the period. There were a range of transformations, some subtle and others more marked, that affected the nature of the power humans and animals were able to wield in relation to each other. Certain kinds of influence, like the emotions animals could elicit and which rested upon heightening compassion for the beasts, intensified in line with a rising popular focus on animal rights and welfare in the postwar period. Similarly, changing technologies meant that people had increasing power to ensure the containment of exotic creatures and to limit their capacity to physically affect human bodies. This chapter draws back the veil on the variety of ways in which captive animals could influence human-animal relationships at the Zoo and how that power relationship was negotiated. It does so by considering three types of influence in turn while recognizing that each of these permeated the entire history of human-animal relationships at the Zoo. I term these three contexts the physical, the reciprocal, and the emotive.

ANIMAL RESISTANCE

One of the most visceral ways in which animals influenced captive environments and crafted human-animal relationships was through physical transgression or the refusal to comply. As described in chapter 1, in the process of acquisition, animals could evade detention or die in the hands of their captors. Once they were at the Zoo, however, captive creatures could alter the human-animal space, influencing the movements of people within and beyond that environment as well as influencing human bodies and minds. The control of space is a fundamental characteristic of social relationships everywhere, the possession of territory usually equating to the possession of power over those within it.[5] Among these kinds of interactions, captive creatures frequently transgressed the spatial boundaries imposed by people.

When animals escaped, human control of territory—and thus governance and spectacle—was challenged. One of the first escapes was of a buffalo that, in 1838, traveled approximately 2.5 kilometers (1.5 miles) to the Bristol city center, where, with keepers in hot pursuit, it was eventually captured and returned to confinement. A few years later in 1846, a leopardess escaped onto the Durdham Downs, a green space adjacent to the Zoo. The appearance of such a threatening beast caused general panic. The *Morning Post* reported that a disabled beggar supposedly found that he had regained full command of his legs, and this allowed him to flee the unsettling scene.[6]

The erection of the Monkey Temple in 1928, already described in chapter 2, might have allowed for the implementation of a barless display rationale, but it also invited the opportunity to escape through human collusion. Twelve rhesus macaques got out of the Temple while it was being cleaned in 1934. Ten of the twelve were captured by nightfall, though only after keepers had spent the day chasing them, not only around the Zoo but also into the surrounding neighborhood. The two remaining monkeys proved challenging to detain. Two masters of Clifton College, a nearby boys' school, and the Zoo's head keeper chased a monkey around a dormitory before managing to catch it. The other monkey had been the subject of "a terrible chase . . . headed by a small boy. . . . [The monkey] would hide in waterspouts and suddenly pop out like a jack-in-the-box; he would scamper lightly along the most precarious ledges and dance among the chimneys."[7] A similar simian exodus of thirty-five monkeys took place in 1949, after three Clifton College boys lowered a ladder into the pit. The Zoo's head keeper spent three weeks reclaiming the animals from the surrounding area.[8]

While the impact of prohibited animal movement illuminates their potential to influence the Zoo as a captive space, cases of unwanted bodily interactions show that animals could quite violently disrupt the Zoo's balance of authority. Through such interactions, animals themselves were able to call into question the Society's ability to control dangerous creatures. Over the past two centuries, the Zoo has been home to violent kangaroos, aggressive llamas, and vicious hamsters.[9] Some violent interactions were especially serious, having far-reaching consequences for the people involved. Great apes, in particular, caused significant physical harm when they violently interacted with keepers or the public during the twentieth century. In midsummer 1938, the chimpanzee Johnny injured keeper Ralph Guise. The chimp reached between the bars, pulling Guise's arm into his enclosure. As watching visitors screamed and a fellow keeper rushed to drive the animal off, Guise's arm was severely torn. Admitted to the Bristol Royal Infirmary for reconstructive surgery, Guise remained there for six weeks before being discharged minus one finger, but at least having secured £300 compensation from the Society for his troubles. While Guise did return to the employ of the Society, he never worked with great apes again, instead taking charge of the birdhouse.[10] In similar fashion, a chimpanzee "savagely mauled" a young boy in 1973. Five hours of surgery were required in order to repair the damage inflicted.[11]

Animals could cause other kinds of harm to people, too. Keeper Jim Chard was particularly unfortunate. He was "trampled" and "savaged" by a zebra in 1962, having had his collarbone fractured when he was thrown against the bars of a cage by the elephant Rosie some years previously. Indeed, as physically powerful creatures, zoo elephants Rajah, Judy, and Rosie were all involved in damaging interactions with both keepers and visitors over the course of the first half of the twentieth century.[12] These breeds of beastly behavior were problematic for the Society. They had the capacity to directly compromise the well-being of the very people for whom the Zoo fundamentally existed, as well as the people who were at the center of its operation.

Such eruptions of animal influence aside, creaturely action could also take the form of smaller-scale acts. These were important moments nonetheless, for they disrupted the husbandry practices designed to intimately manipulate animal bodies. Captive creatures frequently made their own choices about their lives, and this often affected how they were managed. Some were not able to live with each other; some gave birth outside designated spaces; and breeding—and successful rearing—was not a forgone conclusion once animals had been matched up, even in the age of meticulously managed conservation breeding programs.[13] From crocodiles who would not eat, to camels who would not move where keepers wanted them, and polar bears who would not leave their enclosure in order to move to new quarters; from the nonuse of flamingo breeding nests to an elephant's consumption of paint and gorillas dismantling cages, animals contested the management decisions designed to control various aspects of their behavior in captivity.[14]

These potent beasts could physically disrupt the notionally ordered and intensively managed zoo environment. Nevertheless, powerful beasts were not permitted to retain their power without a fight. Human-animal relationships are about negotiation: when animals become too powerful we tend to impose measures on them in order to bring them into line. Thus, physical influence at the Zoo was curtailed in ever more sophisticated ways over the period, while discourse surrounding animal action served to reabsorb undesirable transgressive and obstructive behaviors into changing anthropocentric frameworks of comprehension that fit within the layers of relationships that we have already encountered. Both of these strategies reasserted human authority in ways that waxed and waned over the period.

Not least, transforming physical measures were deployed to prevent escape, limit the amount of contact possible between human and animal, and force captive creatures, as far as possible, to comply with the husbandry practices formulated for them. Over the period, enclosures became increasingly better—some exceptions like the Monkey Temple duly noted—at containing animals, even as the bars came down. The risk of animal violence, meanwhile, could be significantly reduced through outright removal or restriction of their activities. These were typical methods of animal control in zoos more generally.[15] In 1840, for instance, the Society sold its Brahman bull, for it had become a danger to the public. It sold its Sambar deer in 1863, too, in light of its aggressive disposition. When the elephant Rajah became difficult to control in 1925 he was swiftly sold, though he died before he reached his new home in Paris. Much more recently, after the elephant Wendy attacked her keeper during routine care on her increasingly painful feet in early 2002, it was decided that "elephant walks in the Zoo grounds at any time of the day now or in the foreseeable future are completely out of the question. It may never be possible to walk her in the Zoo grounds again."[16] More immediate control of animal action, of course, could be achieved through the provision of firearms.[17] The euthanasia of two polar bears at the outbreak of the Second World War was an explicit attempt to rein in the capacity of dangerous creatures to inflict bodily harm on humans both in and around the Zoo and was an important preemptive measure in anticipation of Luftwaffe raids on Bristol.[18]

Structural modifications and the provision of visitor information were also introduced to reduce the impact of undesirable animal behavior. A sign, for instance, was erected on the front of the lion's cage in 1875, which read: "Notice, Visitors are cautioned not to stand immediately in front of the Lion's den when the animal is moving." In so doing, it reduced the opportunity for unsolicited human-animal contact. By the mid-1970s a number of barriers had been erected following the 1973 chimpanzee attack in order to prevent physical contact, while at least three notices strictly discouraged visitors from touching the animals.[19] In the contemporary Zoo, moats and glass barriers make touching the animals almost impossible in many contexts, thereby limiting the potential for unwanted activity by both animals and their visitors. Just as in other captive contexts, the final quarter of the twentieth century saw far fewer animal attacks (or escapes) than at any other time in the Zoo's history.[20] Intervention was the key, both physically in the form of structure and pedagogically in the practice of instructing visitors about appropriate or safe behaviors.

The physical responses to many of the smaller-scale but nevertheless inconvenient behaviors, meanwhile, were plain and simple patience, increased force, or restrictive measures. A snake, for instance, was force-fed when, in 1929, it could not be gently cajoled into eating its meals.[21] While artificial insemination has not taken place at Bristol Zoo to date, the Royal Zoological Society of Scotland recently used this technique to attempt to help their female giant panda, Tian Tian, conceive, she herself having been uninterested in her male suitor, Yang Guang. This illustrates the ability of humans to try all means of forcing certain behaviors.[22] Contraceptives, however, have been used by the Society since the 1980s in an effort to curtail overproduction (see chapter 1). In the process the control that captive animals have over their own bodies was intimately managed, and there is much potential for even more intimate controls in the years to come.

Physical interventions such as these aside, animal actions were also reconfigured by anthropocentric *interpretation*. Such interpretation was the product, in large part, of interaction between the Society and the press. By communicating their messages through differing channels, altogether reconfigured images of rebellious creatures were crafted. Of course, these configurations may not have been intentional devices through which to wrestle back control of the dominant discourse. Yet, that is indeed one effect that they had. The domesticating anthropomorphic devices, which rose to such heights after the first decades of the twentieth century, were particularly successful in reframing supposedly subversive animal behavior. The monkey who "danced among the chimneys" of Northcote Road, for instance, was transformed into a delightful spectacle. The press reveled in the fact that it had "been the shrine of hundreds of sight-seers who spent hours watching his antics on the house tops." Indeed, the "merry" nature of the event and the "grinning expressions" (in reality an expression of fear, though easily interpreted otherwise) of the animals were accentuated. Keepers were quoted responding in just such an unworried manner. In so doing, the disruption the animals had caused, and the threat they undoubtedly posed, were substantially reduced; their flight from the Zoo was transformed into a theatrical event in itself, which, of course, was very much commensurate with the family-friendly image the Society wished to promulgate for its Zoo.[23] Along the same lines, other twentieth-century absconders were cast as foolish, weak, or ungrateful. They were positioned as vulnerable wards of a benevolent institution. A female raccoon, who escaped in 1937, leaving her young in the enclosure, was portrayed by members of the press as "fickle." Meanwhile, the Society cleansed the animal of such negative portrayals by expressing concern for the welfare of the creature, fearing that she may fall prey to the harm that afflicts all juveniles "when they run away from home."[24]

Conversely, and reflecting the contradictions at the heart of so many human-animal relationships, the explicit deployment of mechanistic language often contributed to the reconfiguration of these animal actions. In response to the monkey exodus of 1935, Superintendent Reginald Greed reassured the public that the animals would be swiftly recaptured. He insisted, however, that this would probably be unnecessary, since the creatures would likely return once they got hungry. In so doing, Greed reiterated the Society's control of its animals, even in unusual circumstances. The implication was that predictable simian instinct would override any supposed desire to continue their adventure. In a similar manner, when an eagle owl returned after a brief unauthorized excursion from its enclosure in 2005, it was purported that "the myth of the wise old owl is just that: they are killing machines who just think about food." Predictable, not uncontrolled, were these creatures; the Zoo was again a home for the animal, a place to which it was *always* going to return. The illusion of human mastery of other animals was maintained.[25]

Similarly, violent acts needed to be understood in the right way. In the first instance, compliance was important if the nineteenth-century Society was to present itself as masterful over the beasts of the wild world and, secondly, if it was to perpetuate its image as a safe place in which families across classes might encounter and understand wild things from the early decades of the twentieth century. Generally speaking, and still today, most people acknowledge that they are predators but find it more challenging to entertain the notion that they might too be prey; animal predation on people shakes the faith that many human cultures have in their implicit right to mastery of the nonhuman world.[26] Subsequently, transgression at the Zoo was greeted with a mixture of horror, understatement, and pragmatic attempts to explain the incident away. The press often responded to animal attacks by evoking notions of a savage and uncontrolled nature at the heart of animal life. A spitting llama, for instance, was portrayed as ferocious and "ill-tempered" in a newspaper report of 1854. One press report in the *Traralgon Record,* relating the elephant Judy's attack on head keeper William Payne in 1931, warned that savageness can sometimes wash away the veneer of tameness among captive creatures.[27] Press reports following the chimpanzee attack on keeper Ralph Guise in 1938 also emphasized the horror of the event, describing screaming visitors and the savagery of the animal in its "maddened" state, rendering Guise helpless at the mercy of the fearsome beast.[28]

By the time of the chimpanzee's attack on Guise, the Zoo had explicitly positioned itself as a place not of wild animals, but of creatures that conformed to expected public behaviors as entities of wholesome amusement and education. The Society thus endeavored to lessen the disturbance to the image of its animals it sought to propagate. In contrast to nineteenth-century guidebook representations, which were often freer with their depictions of wickedness and malice, the animal image of the 1900s was usually publicly sterilized, safeguarding the Society's commercial and financial well-being in a period when zoo animals were presented and perceived more dominantly as pets or friends. A press report of an attack by the elephant Judy in 1931, for instance, featured Society authorities almost laughing the incident off by placing emphasis on the "playful" nature of the animal. In so doing, they sought to limit any scope for the imagination of a darker intent inherent in Judy's unwelcome behavior. The Society similarly rationalized the chimpanzee attack on Guise as most unusual; it was depicted as an all-too-human response to the hot weather (because we all get flustered in the heat, don't we?), or a manifestation of jealousy rooted in the affectionate nature of the creature, since Guise had previously been petting Alfred the gorilla in an adjacent cage. The animal had never behaved in such a manner before, they insisted.[29]

The attribution of blame was also an important facet of the mitigation of these incidents. In the nineteenth century, many species, not only those in captive spaces, were seen as "bad," and bad behavior was simply the result of wicked character. There was no attempt to blame others. Creatures in these contexts were morally culpable but, importantly, and unlike many contexts in circuses and zoos in America especially, there was no retributive punishment.[30] The story was different by the twentieth century, however, when the Zoo was home to animal friends, the cute and the cuddly. In these contexts, blame for animal action was attached to visitors and keepers rather than to the animals themselves. These were the beings conveniently—exclusively—endowed with moral culpability. Therefore, while the intensification of anthropomorphic looking might have widened the moral circle, as described in chapter 4, it remained too narrow to include culpability for transgressions. In 1930, for instance, a boy was found to be "entirely to blame" for an accident in which he was bitten by an otter. Superintendent Reginald Greed similarly blamed a wolf attack in 1952 on a boy who, having had his thumb nibbled, required treatment at the Bristol Royal Infirmary. He had, supposedly, been viewing the animal from an "unsafe" position. The chimpanzee attack in 1973 was similarly attributed to a boy having purposefully reached into its cage. The Society emphasized that it was physically impossible "for chimpanzees to reach out across this eight foot area," which is true, but which nonetheless ignores the fact that an animal bite requires the participation of the animal itself.[31] In the same way, a 1981 crocodile attack was chiefly attributed to the boy involved (and, in a striking parallel to the recent furor revolving around the killing of Cincinnati Zoo's mountain gorilla Harambe, to the sort of parents who turn a blind eye toward unruly juvenile behavior, more generally). All of these animals were decreed to be blameless even though their part in these events was, quite plainly, absolutely pivotal to the unfolding of a damaging interaction.[32]

Similarly, when examples of animal noncompliance were reported across much of the twentieth century, the creatures were presented in ways that construed them as misinformed, amusing, or juvenile. In so doing, their obstructive influence in their captive space was construed in a rather derogatory light. Such representations of problem creatures making light of or belittling their influence were widespread. A camel that would not move toward a waiting train at London's Olympia in 1927 was portrayed as "typically" feminine, insisting on having the last word rather than knowing what was best for her. A young crocodile, which, in 1929, would not eat, was reported to be showing a distinct lack of appreciation for his new home, and, during the same year, it was conveyed that snakes had to be force-fed until they "realised the error of their ways." More recently, the elephant Wendy was considered "very naughty" when she refused to go back into her house at nighttime.[33]

RECIPROCAL RELATIONSHIPS

These influential acts reveal that many kinds of relationships with animal life at the Zoo were not always a series of one-sided human manipulations but, instead, frequently involved processes of interspecies negotiation where the balance of power perpetually swung back and forth. Within this interactive environment, interspecies cooperation produced some of the foundational relationships between specific human and nonhuman animals over the span of the period. Feeding, petting, and interactions between keepers and kept were not only imaginative affairs but were also, in fact, mutually constituted.

While it was explained in chapter 2 that the visitor feeding of animals had been central, since the nineteenth century, to a day out at the Zoo, food was not simply shoveled into the gaping mouths of passive creatures. In fact, between its introduction in the 1840s and its near-prohibition in the early 1960s, the process of visitor feeding was very much a coproduction resting on reciprocal social behaviors. Many creatures, especially mammals and birds, developed behaviors through which they actively sought food from visitors. Indeed, the bears' climbing of their pole in order to receive foodstuffs was itself a specific exhibition of *interactive* behavior made by both humans and animals. The bears were not forced to ascend; their behavior instead became conditioned in anticipation of feeding, and this feeding could only take place with the cooperation of the climbing bears themselves. Primates tend to be particularly adept at coaxing food from the hands of their humans. It was noted in 1931, for instance, that apes "are very fond of walnuts. As soon as they spy their keeper [or visitors], outstretched palms are thrust forward" (fig. 26).[34] Seals similarly developed such behaviors (fig. 27). In so doing, habitual human and animal behaviors were reinforced and physical thresholds crossed.

Similarly, the provision of an animated spectacle more generally, and specifically acts such as riding, rested upon animals playing their parts. Yet such

FIGURE 26. Adam the chimpanzee, c. 1936. Rodney H. Webb. (Bristol Zoological Society Archive, courtesy of the Bristol Zoological Society)

physical exchanges were themselves subject to interpretation according to anthropocentric concerns. They never stood alone as moments of physical interaction but were remolded to present the animals as performers, existing predominantly for the pleasure of people. This was the case throughout the period, but it intensified when representational devices of the twentieth century positioned animals as desperate for human attention. The Society, the press, and the public conspired in the promotion of this kind of interpretation of human-animal interaction. The elephant Rosie, for instance, was portrayed as "never happier" than when she was serving the Society by giving rides to children, while monkeys "enjoyed the company of a big audience."[35] In 1977, one journalist wrote: "All zoos have their comedians and Bristol Zoo is no exception. One of theirs is Sasha the seal who is always ready to extend a welcome flipper. Sasha patiently waits for a crowd to gather, then stands up with flippers resting on the wire barrier. . . . Sasha slyly lowers one flipper and scoops up enough water in one powerful movement to drench them."[36] The

FIGURE 27. Feeding sea lion, c. 1929. (Roy Vaughan Collection, Bristol Zoological Society Archive, courtesy of the Bristol Zoological Society)

physical engagements between visitors and the elephant Rosie, the monkeys, and Sasha the seal were, in these contexts, represented within an anthropocentric frame. The creatures apparently reveled in the entertainment of their visitors. That was, after all, what they were there for. Indeed, some reporters wondered who was watching whom, suggesting that the animals were as much amused by their human admirers as the animal spectacular entertained visitors.[37]

Equally mutually shaped were the relationships that developed between particular people and the animals they managed. Thomas Nagel prompts us, in his ruminations about what it might be like to be a bat, to consider that the actuality of these kinds of relationships are impossible to fully discern without access to the inner worlds of the animals themselves. Indeed, such worlds were somewhat occluded by anthropomorphic projection. Reciprocal relationships, however, have always been a way in which zoo animals could craft the intimacies of captive space.[38] Elephants, for instance, appear to have formed strong bonds with particular keepers. Indeed, the strength of these relationships is unsurprising, since they echo those between elephants and their mahouts. In nineteenth- and early twentieth-century British zoos, it was standard practice for an elephant to be accompanied on the journey from Southeast Asia by its mahout, who would subsequently assist in the handover of care. Mahouts in India (and oozies in Burma) are assigned elephants early in life, and the bond between them is often life-long.[39] Rajah, for instance, had an affinity for his keeper, William Payne. Payne joined the Army Veterinary Corp in 1917. Upon returning on leave, he visited the elephant house. He recalled, "although I was in uniform and he had not seen me for many months Rajah remembered me. He caressed me with his trunk." Payne never returned to the Society's employ full time, and Rajah swiftly became intractable. No longer trusted to give rides to children, the Society made arrangements for him to be sold to another zoo in 1925, and a replacement riding elephant, Judy, was duly acquired to take his place.[40]

The elephant Rosie exhibited a similar degree of attachment to her keeper, Tom Bartlett. Rosie, acquired to replace Judy in 1938, and Bartlett arrived at the Zoo trunk in arm. They had been together for some years previously during Rosie's career as a circus elephant, and they were supposedly inseparable. So strong was the bond that when keeper Jim Chard took over her management during one of Bartlett's rare vacations in 1941, he was attacked and his collar bone broken. Bartlett was immediately recalled to the Zoo and ordered to resume charge of the elephant. No more holidays for a keeper whose working hours already well exceeded the forty-odd hour week of the average worker in the 1950s. Some years later, in 1958, the traditional spectacle of a ride on Rosie was postponed. Bartlett had been too unwell to train her before the Easter holidays. He was getting older, and the Society's President of Council John Spencer Young was becoming increasingly anxious about the consequences should anything happen to Bartlett before a replacement could be identified and then accepted by Rosie. The episode with Chard had shown that Bartlett would not be easily replaced. Young's concerns were well founded: Bartlett retired in 1959, and Rosie died a mere eighteen months later. Staff members were of the opinion that she had fallen victim to a broken heart.[41]

These kinds of interspecies scenarios reveal, at the very least, the possibility of a close attachment between keepers and kept that significantly influenced the way the Zoo was managed. Given the wider evidence of attachment between elephants and their mahouts in the animals' native spaces, it is reasonable to read some degree of reciprocity into the interspecies relationships unfolding here. In other words, this was not mere projection.[42] As described in chapter 4, close relationships between keepers and kept pervade the entire span of the Zoo's history. Of the close interspecies interactions examined there, those depicted by former senior keeper Mike Colbourne were among the most intense. The ways in which he recounted his bond with the great apes he managed were not—if we consider them in the ways we considered the human-elephant relationships above—fanciful delusions. Remember that Colbourne spent a lot of time inside the enclosures of his animals, relating to them in very proximal, physical ways. Ethologist Heini Hediger postulated that in zoos, "the keeper is only *allowed* to enter cages where the inmates recognize him as a socially superior member of the same species."[43] Colbourne recalled just such mechanisms of negotiation and recognition. He suggested that a set of rules had been mutually generated: "Gorillas are majestic, intelligent and fascinating and above all they respect fair play. I have to dominate them and earn their respect, and we have agreed that a certain tone of voice on my part exacts obedience and a certain noise and facial expression on their parts means they have had enough of games and my company and want me to leave the cage. Without these rules I could be hurt."[44]

Colbourne employed a number of signs to facilitate communication between him and the gorillas. This is hardly unusual: signs were important communication tools in the infamous late nineteenth-century Clever Hans phenomenon, for instance, in which an Orlov Trotter horse was supposedly able to carry out simple arithmetic. It was later found that the animal was watching and responding to the subtle body language of his trainers. The gorilla Diana, when taken for a walk in the Gardens, would pull on Colbourne's hand and lead him back to the enclosure when she had had enough of her foray among her adoring visitors. Moreover, Colbourne noted that the gorilla Delilah, after permitting him to sit with her and her newborn baby, Daniel, in their enclosure in 1971, eventually allowed him, on the fourth day, to hold the young gorilla. He also claimed to be able to speak "gorilla," and that he utilized this skill in understanding the animals and communicating with them.[45] The degree of touch evident in Colbourne's relationships with these animals and the utilization of communicative signs at least suggests some degree of negotiation between human and nonhuman individuals in the production of a relationship, with the gorillas being physically strong enough to violently reject interaction if they so desire.

Once again, however, the influence of these animals on the manufacture of human-animal relationships was to some degree distorted through anthropocentric interpretation. The press's depiction of Colbourne's position as "mother" to some of the great apes clouded the reality of the interspecies interactions unfolding.[46] Reciprocal relationships, then, existed at the Zoo across the period, and while their realities are clouded, we can certainly chase the shadows cast on the walls of this institution of captivity.

POWER FROM A PIT OF DESPAIR

All of this has illustrated that captive animals interact with those around them in a diversity of ways and in so doing impact upon a number of layers of relationship. Importantly, however, neither animals nor the humans with whom they immediately interacted existed in a vacuum. They were born of their time and were held captive in it too. Sure, they could bite, run, and reciprocate, and these behaviors could be managed and interpreted in a variety of ways over the course of the period. However, these moments tend be just that: *moments in micro-space.* All beings exist in connection with each other and with the circumstances of their time. In fact, it is in the context of this kind of sociopolitical "ecosystem" that captive animals became ever more powerful as cultural influences over the course of the period. As public attitudes to wildlife transformed, particularly in the postwar world, captive animals, and some in particular, came to be influential in the mobilization of that sentiment into meaningful sociopolitical action. As sentimental attitudes toward animals intensified, the image of animals as victims of domination became increasingly entwined with wider cultural attitudes around exploitation, animal life, their captive spaces, and the wider world.

In order to illustrate both the capacity of animals to be influential in wider sociopolitical systems and the ways in which this changed over time, I turn to one particular creature whose distinctive presence helped to turn the fortunes of the Society on its head at the end of the twentieth century. Importantly, his influence was unquestionably a product of his time. In the late 1980s, the Society received a letter from a member of the public. It read "please send that poor polar bear back to freedom in the Arctic, or God will [punish] you with hot coals on your head." At the same time, the Society received letters from visitors to its zoo, acknowledging what they saw as effective captive care of the creatures of the world's ever more fragile wild places, and supporting it in the face of what had become a relentless public onslaught over the welfare of its increasingly infamous polar bears.[47] Over a decade earlier, in 1979, the Society had acquired an adult male polar bear named Mishka. Bought from Chipperfield's Circus, his arrival followed an extensive search for a male to replace the Zoo's deceased favorite, Sebastian.[48] By 1985 the bear, renamed Misha, was one of three animals occupying the naturalistic ice-floe exhibit in the southwest corner of the Zoo. While he may have been a specimen of an enduringly popular captive species, Misha was a bear who behaved in ways that the public and the Society increasingly judged to be alarming. He endlessly paced his enclosure, four steps forward followed by four steps back, swaying his head repetitively from side to side as he strode. His apparently psychotic appearance in a captive environment evoked images of the prisons and asylums of the past, where surveillance, spectacle, and suffering collided to alter the natures of the beasts within. After observing Misha in his enclosure in 1986, Zoo Check, which had formed in 1984 in response to the plight of the elephant Pole-Pole, who likewise exhibited signs of mental disturbance at London Zoo, and which later came to be known as the Born Free Foundation, published the *Captive Polar Bears in the UK and Ireland Zoochotic Report*. The report found that a majority of captive polar bears in the UK and Ireland, a species so habituated to ranging vast distances in indigenous environments, exhibited behavior indicative of poor mental health in response to their confinement. Their use of the word "zoochotic" deliberately evoked visions of frightening psychoses and disturbing abnormality.[49] Among these polar bears, Misha was said to be displaying some of the most severe stereotyped characteristics. He was the poster boy for anthropogenic insanity.

Stereotyped behaviors among captive animals are not unusual.[50] Without a doubt, polar bears had exhibited such mannerisms at the Zoo before. Visitors in 1889 observed pacing and head swaying, with one report stating that "these bears, like others of their kind, are fond of walking up to the bars of their cage, and then retreating backwards, wagging their heads from side to side; and one of them has the curious habit of walking through the water to one end of his tank and then swimming to the other end on his back."[51] Similar observations were made well into the mid-1930s.[52] Yet these animals of an earlier time were not influential in the manner of Misha. Their behaviors were curious, but harmless.

Misha's behavior was different. It came in the last quarter of the twentieth century, when the public gaze had decidedly turned toward the sentience and suffering of animals in a mechanized, utilitarian society. The anthropomorphic and increasingly emotional tendencies of much of the previous half century or more had taken their toll, permanently reconfiguring attitudes to animals in light of their ability to suffer *like us* and in the context of a powerful cultural sensitivity to domination and inequality. Propelled within a receptive cultural landscape, the *Zoochotic Report* generated significant press coverage focusing on the plight of the bear and, to a somewhat lesser extent, his companions, Nina and Janina, who exhibited milder forms of stereotypy.[53] Headlines such as "The Sad Bear of Bristol Zoo" and "Why We Must Stop This Madness," referring both to the "madness" of the bears and to the "madness" of their being kept confined, fixed public attention on the conditions in which large animals were maintained in both Bristol's and the nation's zoos. These places, for so long at the jovial core of the nation's downtime culture, became synonymous, in both the local and national press at least, with animal pain.[54] Letters did not simply object to the captivity of polar bears but to the entire notion of artificial animal environments altogether. One correspondent wrote to the *Bristol Journal* noting that "with marvellous nature programmes on television the need for these Victorian monstrosities to satisfy human curiosity must surely be diminishing."[55] After a while, the ferocious criticism subsided, but it surfaced again in 1989 when Bristol city councilors renewed the Zoo's license allegedly without having

considered the captive conditions of the polar bears. The bears—"doomed,"
"mental cripples"—were portrayed as dwelling in a "pit of despair," waiting
in a state of abject suffering for the merciful release of death.[56]

Over three hundred letters, from all parts of the United Kingdom, poured
in to the Zoo between 1986 and 1989. Many of the letters referred to the
misery of Misha's life, suggesting that it would be more humane to kill him
outright than to allow him to endure, while a large proportion suggested that
he should be returned to what they perceived to be some kind of idyllic freedom in the Arctic.[57] "How could you do such a despicable thing," one woman
demanded to know, "a far cry from the image you projected with Johnny
Morris," referring to the heavily domesticated and sentimental world of BBC
television's *Animal Magic.* Another correspondent wrote to the Zoo's director, seething: "Mr GREED. You certainly have a very appropriate surname,
though I prefer to call you . . . something else, an unfeeling bastard over your
attitudes to the plight of the polar bears. I hope nobody even visits your rot-
ten, poxy, zoo." Misha's behaviors also had a profound effect on children, many
of whom wrote to accuse the Society of cruelty. They implored it to return the
animals to the idealized sanctum of the wild places.[58] During the period of
what I want to call the Misha Crisis, and when its fallout was most clearly positioned in the public domain, visitor numbers fell rapidly to their lowest ebb
since the 1940s.[59] As Zoo Check noted in 1989, "there can be hardly anybody
in the country who has not heard of Misha the pacing bear at Bristol Zoo."[60]

Yet, while this furor might have been a major problem for the Zoo and its
public image, it actually inspired important innovations in animal husbandry,
showing how Misha was influential well beyond the evocation of outrage and
humanitarian sentiment. A campaign was launched by the *Mail on Sunday,*
which raised some £12,000 in public donations that were then put toward
improving the lives of captive polar bears throughout the United Kingdom.
A package of ethological studies was administered by the Universities Federation for Animal Welfare, which indicated that changes to diet and the
enrichment of captive environments through strategies to mentally stimulate
captive animals by promoting play and foraging activities could significantly
reduce displays of stereotyped behaviors. The implementation of these sug-
gestions led to a significant decrease in stereotypy among Bristol's polar bears.
This received widely positive reportage in the national press, and this in turn
inspired pervasive implementation in zoos throughout the country.[61]

Misha's behavior thus had significant consequences for human-animal relationships in the context of the lives of captive animals themselves, and this influence extended well beyond the immediate microcosm of Bristol Zoo. Importantly too, however, Misha's disturbing—almost anthropogenic—behavior drew attention to the plight of other captive polar bears, including the enclosure conditions endured by the popular London Zoo bear Pipaluk, who had been residing in a "grim and lonely" cage in Poland since he was sold in 1985.[62] The fundamental rationale underpinning the existence of zoos more generally was also called into question. Press articles in the late 1980s and early 1990s, often taking the *Zoochotic Report* as a starting point, scrutinized the continuing role of zoos. Many concluded that essential changes were needed in order to reposition zoos in relation to a worldwide conservation mission. Only this, they determined, could justify the existence of such relics of the Victorian age.[63] In Bristol, the work of Zoo Check, continuing adverse publicity, and the enduring public antagonism directed toward the Zoo accelerated a fundamental change in the Society's ethos that had become essential to the health of its public image in this period of increasing ecological awareness and re-evaluation of zoo missions. Subsequently, and as we saw in chapter 1, the Society accelerated the dispersal of its large animal collection to zoos with more space to devote to enriching captive lives.[64]

Misha was euthanized in 1992, immediately following the death of his companion Nina, who had also been euthanized after the onset of liver disease. It was decided that Misha's life would have been "intolerable" without his constant companion and might have precipitated a relapse into stereotypy to the disturbing degree of the mid-late 1980s.[65] The bodily departure of the image of the "mad" polar bear was far from the end of the matter, however. The disturbing specter of Misha could not be erased. His influence endured in the form of a local protest group, Bristol Zoo Action, which focused its attention on the lives of other large animals at the Zoo by picketing the main gates during the summer of 1993.[66] Their protests, inspired by the controversy surrounding Misha, concentrated on the Zoo as an "animal prison," incarcerating freedom-loving creatures, particularly the Sumatran tigers, which now occupied the old polar bear enclosure, and the Zoo's solitary—and last—elephant, Wendy. They lobbied for her immediate removal from the Zoo, claiming, without any explicit evidence, that she spent most of her time chained to the floor of her enclosure.[67]

Alongside the considerable response to Misha's behavior in an increasingly unpalatable artificial environment, the affair also served to pique concerns about the state of the natural world in the shadow of accelerating ozone depletion and the retreat of the continental ice shelf in the ecologically fragile frozen North. In 1989, *Punch* magazine criticized Zoo Check for endorsing the closure of all zoos without any consideration of escalating global conservation issues. Pointedly identifying the scientific and conservation issues lurking behind popular images of the cute and cuddly victims of nefarious zoo oppressions, *Punch* remarked that "Misha the polar bear used to live in a circus cage. It sent him mad. He spent ten years recovering from his experience in Bristol Zoo. Then animal rights group Zoo Check discovered him. Now he's got two personal psychologists, hundreds of concerned 'friends' and a Polar Bear Action Account at the Abbey National. It will probably send him bananas again."[68] Bristol-based Aardman Animations' stop-motion *Creature Comforts* (1991) likewise recognized the depleted state of the wild worlds of the North. Zoos are like "nursing homes" for animals, a plasticine juvenile polar bear announced, implying that the creatures were increasingly unable to live independently in their own "homes."[69]

In these ways, a number of attitudes toward the rest of the natural world, at once engaging with utopian and dystopian visions of the frozen North and the broader state of the planet, revolved around the disturbing forms of Bristol's polar bears. In a Foucauldian sense, Misha's "doomed" body was a contested site, a vehicle of power where authority was enacted and resisted.[70] Misha helped to inspire, was swept up in, and became emblematic of a gathering concern for both the welfare of animals and the state of the planet toward the end of the twentieth century. While his body operated in expansive biopolitical arrangements and contributed to significant ideological shifts, his influence was in large part a product of his time. His physical movements and the image they evoked were influential to the deepening of a cultural milieu that reflected changing symbolism attached to what is now a species emblematic of a damaged planet.[71]

Despite all of this, however, and like other moments of animal influence at the Zoo, the power of Misha's body was unstable and subject to dilution, if not total elimination. He was, both physically and figuratively, the subject of rigorous management. As we have seen, press and public responses to the bear were characterized by anthropocentrism. Perceptions of sadness and misery detailed above were unverifiable evaluations of the mental worlds of the creatures that were, perhaps, indicative more of the self-centered feelings people had toward the animals and less of how the creatures genuinely felt in their "pit of despair." The Society itself tried desperately to control the public impact of Misha's behavior in a number of ways over the period. Not least, the sight of Misha's disturbing behavior was censored from public view in 1986, when he was moved into an off-show area and the sign featuring his name removed, all but erasing him from the Zoo's animal spectacular. The Society also proclaimed, along with Glasgow Zoo and Flamingoland, that they would no longer be keeping polar bears after the current crop had departed. In so doing, they tried to ward off any escalation of hostilities.[72] Similarly, Misha's euthanasia when he was in an otherwise good state of physical health might be considered the ultimate preemptive elimination of animal influence, given that the Society was fearful that he would relapse into stereotypy in the absence of his long-time partner, Nina.

These acts of censorship embodied a strategy long employed by the Society in response to unusual animal behaviors, though in Misha's case these measures were of extraordinary importance in an age when animal misery was a potent influence in the crafting of the public perception of zoos. In 1935, for instance, the Society responded to complaints about an orangutan by euthanizing the animal.[73] Whatever the nature of the animal's behavior, it was evidently considered to be unfavorable enough to merit eradication from a family-friendly public display.

In addition to this brand of physical censorship, the Society also resorted to intense management of its public relations amid the hostile press relations of the 1980s and early 1990s. In an age of ever expanding mass media, controlling opinions and attitudes toward the animal spectacular was vital. Director Geoffrey Greed's reaction to a BBC Radio 4 interview by Julian Hector, former scientific adviser at the Zoo, who at that time worked for the Wildfowl and Wetlands Trust at Slimbridge, just north of Bristol, is telling. Greed was most upset at negative comments made on air. The remarks, he noted, were not at all good for the Bristol cause. As a result, Hector was asked, and agreed, not to discuss Bristol in future broadcasts relating to animal behavior. Later, a press release was sent to the editors of a wide range of publications following the second wave of reports in 1989, looking to set the record straight vis-à-vis measures that had been put in place to enhance and enrich the captive lives of the animals.[74] It is no coincidence, too, that the first member of staff responsible for managing public relations was employed in 1990, in the immediate aftermath of the negative and often aggressive press coverage of the previous years. Indeed, a number of zoos reacted to the Misha Crisis and its fallout by being more proactive in pushing positive news stories into the public domain, rather than simply being reactive to negative coverage. Such measures equipped the Society with the tools it needed to successfully defend itself in an era of heightened sensitivity to the fragility of captive lives. Stereotypy displayed by the Zoo's Persian leopards in 1993, and caught on camera, was thus easily dismissed as the cats merely "reacting to the cameras."[75]

Of course, this is not to say that such anxieties regarding the management of public relations are a very recent phenomenon. The Society's image has always been at the heart of its success, and from early on, good management of press relations was essential. The General Committee expressed concern over the leaking of a lion escape in 1866, since it brought into question their capacity to control their animal wards. In addition, following reports on the elephant Judy's attack on head keeper William Payne in 1931, Payne was warned to be more careful in the future about the sort of information that was freely given to the press. These kinds of procedures, however, are more vital today than they were in previous periods. The public became so sensitive to some forms of animal exploitation in the second half of the twentieth century, and the zoo such an increasingly tainted institution, that its very survival became dependent on rigid control of the shadows captive animals cast on the world beyond its walls. Indeed, the Society today possesses a "whistle blowing policy," which details the disciplinary action that might be taken if any unauthorized employee were to talk to the media.[76]

Whether the events and conversations encircling Misha's body in the 1980s and 1990s might have come about without the unintentional actions of this specific captive animal is, perhaps, without consequence. What matters is that Misha's body and behaviors were intricately bound up in a vast web of actors and human-animal relationships positioned in space and time. Through the relationships between the animal and the people and places surrounding it, he was able to contribute to environmental and ideological change within and beyond the site of his own captivity. Despite our lack of access to the realities of Misha's mental worlds, he was clearly an individual fundamental to the physical and imaginary complexion of a range of hybrid historical environments and mentalities. While similar behaviors among the polar bears of 1889 and 1894 triggered curiosity among those who viewed them, the late twentieth-century sociopolitical climate positioned Misha as a historical being in a significant biopolitical network. His influence sits firmly within a very specific historical frame in which the troubles of both captive and wild creatures were able to elicit powerful empathetic responses not only from animal rights advocates but also from the public at large. Yet his influence was not allowed to stand for itself. Like escapes, attacks, and various other forms of resistance and relationship, Misha's body was a site of conflict and a locus around which the power of the human and nonhuman competed for control of the message leaking between the bricks of the zoo's boundaries. This variety of human-animal relationship is not about monolithic or even subverted power relationships; it is about authority forever in flux. Just as an ecosystem pulses with the interactions of lively forms, so too did—and does—the zoo. Made and remade again and again, it is often a product of humans and nonhumans entwined, but the precise nature of that interaction, of the human-animal relationships within, retains its roots in time.

CONCLUSION

The human domination of other animals, so often perceived to sit at the very heart of the zoo as an institution, was far from absolute. Animals had the capacity to exert influence over environments, bodies, and minds in a variety of contexts. Through interactions that had both physical and emotional impacts, captive creatures were powerful beasts, their interfaces with humans at the Bristol Zoo, and sometimes far beyond, often the subject of negotiation. The ways in which animals affected change actually influenced—and were influenced by—all of the layers that we have already identified, from the death of a valuable lively commodity to the ability of an animal, by virtue of its morphology, to elicit affection. These actions, however, were, like every other aspect of their relationship with humans, subject to interpretation. Deliberate attempts by the Society to physically curtail animal action sat alongside less explicit, though no less effective, modes of enforcing the provision of an amenable and compliant animal spectacle through managed bodies and representations. Interpretations of unusual or unexpected animal behaviors on the part of the press and the public served to construe animal actions in a way that reframed animal influence according to human preoccupations and assumptions. Though the nature of this recolonization of the influential animal form changed in many ways during the course of the Zoo's history, it remained effective in obscuring the authenticity of animal action throughout.

Attempts to reduce the impact of instances of undesirable performance, whether attacks or behavioral indicators of psychological disturbance, reveal that animals were supposed to maintain a persona; they were supposed to remain within the confines of the layer—or layers—of relationship within which they were supposedly captive. By the twentieth century, this persona was sanitized, unthreatening, and certainly not disturbing, and it was commensurate with the Society's and the public's ideas about what captive creatures ought to be. In the combination of animal action and its perpetual interpretation, human-animal relationships at the Zoo were forever in flux, perpetually embroiled in processes of meaning making.

6

The Faces of the Beasts

Few tourist attractions can compete with zoos, aquariums, and wildlife parks in terms of their public appeal. An estimated 700 million people worldwide visit animal spaces linked to the World Association of Zoos and Aquariums (WAZA) alone every year. Such is the magnetic attraction of exotic worlds writ small. When we visit these places, we might think that we are heading for an extraordinary encounter. In reality, though, zoos are, in some important ways, microcosms of our everyday encounters with nonhuman animals.[1] Zoos have proved themselves to be unstable places, their multispecies complexions kaleidoscopic in nature. They craft, reflect, and react to colorful cultural understandings of the more-than-human world. Thus, across nearly two hundred years since the founding of the Bristol, Clifton and West of England Zoological Society in 1835 and the opening of its zoo to the public the following summer, the character of human relationships with captive creatures at Bristol Zoo has seldom been simplistic or static. At any single point in time, there existed an astonishing array of attitudes toward animals and a perhaps surprising turbulence to the tides of power that surged among species. The traditions of relating to animals described over the course of this book might give the impression that an animal could be seen, in various degrees, as a commodity, or a pet; an amusement or an object of research, proximal or distant; an individual emotional being or a specimen emblematic of its species. But the imagination of captive animals—and indeed our understandings and thus treatment of animals more generally—was composed of layers that existed simultaneously, one on top of the next, forming a rich world of human-animal relationship. People—the Society, its visitors, and the media—could possess any number of these apparently contradictory attitudes to the animals they encountered *all at once.* As Eric Baratay and Elizabeth Hardouin-Fugier so eloquently put it, in the zoo "every aspect of humanity's relationship with nature can be perceived . . . repulsion and fascination; the impulse to appropriate, master and understand; the progressive recognition of the complexity and specificity of the diverse forms of life."[2]

Through commoditization into objects of various extrinsic, intrinsic, and species-specific values, the animals that flowed into the Zoo were transformed into a magnificent and multisensory spectacle whose purpose was to amuse and inform in various ways across the nineteenth and twentieth centuries. They were imagined as scientific specimens in whose bodies an ever-deepening repository of knowledge could be found, and as almost-persons whose amiability rose in intensity before declining, even if only in small measure, across the second half of the twentieth century. Importantly, too, animals actively influenced the creation of their captive space and the wider cultural world far beyond. Through our journey among these layers of human-animal relationships, the natures of the beasts have been shown to be many and complex. These identities, usually anthropogenic, have always been rooted in a binary at the heart of human understandings of animal life: that of the object and the subject, and the tension between the two causes confusion.

These layers of relationship took on particular complexions at Bristol Zoo as a special kind of captive space that itself was perpetually made and remade. Throughout the Zoo's history, animals were conceived of as objects of varying commercial and scientific values. Many were prized because of their significance in furthering the zoological ends of the Society, whether that be the desire to forge a comprehensive collection in the nineteenth and early twentieth centuries, or in the later context of the reification of precious genetic material. Others were valued, right across the period, in the contexts of their sex, age, rarity, or ability to generate and entertain crowds whose pockets were lined with silver. Further objectification took place in the crafting of amusements of an often multisensory character across the entire period. Captive creatures were also objectified as scientific specimens whose bodies could be examined in ever-greater depth and detail. This knowledge was put to good use, not only in innovation in captive conditions and husbandry techniques but also in the gathering of understandings about the animal occupants of the rest of the natural world. Yet, in the transformation of animals into almost-humans—whether the antagonistic animal foes of the nineteenth century, the animal friends of the twentieth, or the enduringly recognizable emblems of fragile wildness at the turn of the twenty-first—they were recognized as subjects who had vibrant lives of their own beyond the contexts of their associations with humans. The mourning and remembrance of captive creatures, and a rising visitor concern about captive conditions, signal the presence of a strengthening view of animals as beings who not only were emblematic of their species but who also were presences capable of forging their own paths through the urban jungle. This recognition happily sat alongside their standing as indistinct representatives of species and as malleable objects of endless exploitation.

This duality of object-subject at the heart of these perceptions of captive creatures reflects a fundamental principle in many human relationships with animals. This way of looking renders the animal and its world simultaneously distant and proximal to the human animal, reflecting some of the anxieties at the core of our status as earthly beings. In the nineteenth century these anxieties were rooted in an uncomfortable resemblance between humans and other primates, for instance, but by the twentieth century it increasingly could be found in the belief that while most of us recognized that we were animal—the animal peering back at us via the mirror of nature-culture emphatically told us so—there must be *something* about us that remains exceptional. Such ambiguous feelings have promoted the drawing in and pushing back of animals and animality from ourselves and from our culture and, in the process, have granted moral permission for the development of relationships rooted in simultaneous exploitation and intimacy.[3] Indeed, this underpins the character of human-animal relationships in captive contexts. Zoo animals were implicitly characterized in this way. They were largely unlike us in their consideration as commodities, as objects of amusement, and as scientific specimens. In contrast, their closeness to the human animal was accentuated in the recognition of their individuality, as beings with highly developed emotional lives and personalities. Yet, when we think back over the layers of relationship that we have encountered and taken apart, it becomes clear that the precise balance of sameness and difference was dependent on historical context, species, and individual experience.

While this book has centered its gaze primarily on themes rather than chronology, if we look back across the layers of relationship encountered here, we can see that the layered nature of human-animal relationships did reconfigure at watershed moments. At these points, human ideas about non-human animals underwent significant transformation. With roots in early nineteenth-century endeavors to collect and classify the human and more-than-human worlds, the Society was initially founded to create an exhaustive living collection of zoological beings. In this context, the animal object provided "rational recreation" for the people of Bristol and allowed for the accumulation and dissemination of zoological knowledge. Animals were typically displayed in small, generally bare cages, which placed them as objects of amusement and instruction at the very center of the curious gaze.

In this period, animals were also often described in human terms, but their difference from humans was nonetheless accentuated through emphasis on disagreeable traits such as malevolence and greed and a certain lack of compliance with the high-minded desires of the human animal on the inexorable long march of civilization. This sense of human separation from the rest of Creation was at its most profound in the years before Darwin's theses on the mechanisms through which species emerged and transformed over time significantly muddied the deep waters of the human-animal borderlands. As already noted, Darwin's insights took a while to permeate, and consequently these ways of looking at animals remained dominant until well into the twentieth century.

As the perception of kinship between humans and their animal others intensified, a tendency to accentuate the agreeable individuality of animals in captivity became more and more potent. Animal personalities were being depicted as ever more affable by the late-nineteenth and early-twentieth centuries, reaching a climax between the 1930s and 1950s. During this period, children's literature and film, in particular, deepened attitudes toward the nonhuman, which were increasingly rooted in a sense of similarity, common ground, interspecies communion, even friendship and love. Human relationships with the famed gorilla Alfred and the chimpanzees Adam and Timothy, for instance, epitomized this intensifying sense of intimacy across the human-animal borderlands. Increasing concern for captive lives and a mounting anxiety about animal freedom and happiness reflects an important change in the complexion of the object-subject binary in this regard. Yet, at the same time, conceptualizations of animals as objects of science deepened and changed course. The science of ethology cast an increasingly intensive gaze onto the forms of moving animals, while the formal accumulation of zoological knowledge meant that supposedly predictable bodies could be managed ever more effectively.

Although the precise nature of the object-subject binary was rarely static, a radical reconfiguration took place from the 1980s, when individual captive creatures were increasingly objectified as emblematic of species in peril, which encouraged humans to cherish genetic material and the bodies that carried it as objects of intrinsic worth. Ways of looking that were prevalent across the middle of the twentieth century became less acceptable, though they retained some appeal through their promotion of a sense of visitor empathy toward the imperiled *individual* creatures on display. These animals became emblematic of a mounting crisis rooted not just in the faraway wilds of the world but also one that encompasses all of us, not only in terms of location but also in the context of our shared experiences of *being* in the world. The new ways in which many animals were displayed and represented enshrined an immersion in a shared world that was innately fragile in nature, though the veneer of commonality to some degree masked the enduring truth that animals at the Zoo remained, at their core, objects of desire, display, and, ultimately, manipulation that was often complicated by benevolent ends.

Despite such changes in the ways in which these layers interacted and transformed over time, there were nonetheless important continuities across the period. We should not shy away from this. Historians of whatever hue tend to be drawn to change, but sometimes an obsession with transformation can obscure the presence of continuities that are both surprising and disquieting. In some ways, animals were treated in the same way at the beginning of the twenty-first century as they were at the beginning of the nineteenth. Despite changes over time, they were consistently objectified within economic systems, either in the wild animal trade itself or through their deployment in the generation of revenue and in systems of knowledge acquisition. They also continued to function as objects of spectacle in a shifting zooscape because the Zoo has remained, in essence, a tourist attraction reliant on visitor-generated income across the entire period. If it were otherwise, if animals were in no way regarded as amusements capable of imploring visitors to part with their hard-earned money, the Zoo would have shut its gates long ago.

Importantly, too, the heart of human-animal relationships in captivity has always been about power. It is imperative to remember that these are not animals, for the most part, making their own choices in life. They have all been moved around the world between wild and captive spaces without their consent, and they have all been placed in environments designed, in substantial part, to enhance their spectacular natures. They have been studied, handled, and even saved, for the ethos at the heart of conservationism rests on the human ability to manipulate animal life and their local ecosystems, even if for benevolent reasons. While captive animals were conceived of according to a vast array of human preoccupations, to suggest that an animal is little more than an amalgamation of our ideas about it and perpetually at the absolute mercy of our manipulations risks rendering it mute, merely peering in from the edges of the world. Traditionally, the zoo has been rather simplistically cast as a place of total human dominion over the natural world. It has been depicted as one among many icons of oppression: the ghetto, shantytown, concentration camp, prison, and madhouse.[4] Yet manipulation and marginalization fail to tell the whole story, for we know that these animals were, in some ways, influential. They frequently cocreated their captive spaces through their behaviors, they forged relationships through the power of their own personalities, and they influenced attitudes toward themselves, their captive worlds, and their wild kin. The Zoo was a space of multispecies making.

Aside from historical continuity and change, the precise nature of the object-subject binary might be reduced down to the level of species and, even further, to that of the individual. Human-animal relationships varied enormously according to the identity of the creature concerned, and this variability, rooted in many ways in the presence of an anthropogenic hierarchy of life, persisted throughout the period. While all animals were conceived of as objects of value, spectacle, or scientific understanding, some were more highly prized than others were. Elephants as beasts of burden, other large mammals, and exotic novelties were particularly esteemed as objects of science or spectacle, and they became even further esteemed when their individuality was allowed to flourish. Indeed, mammals, and great apes in particular, were eminently capable of complicating the already complex object-subject binary, blurring the human-animal boundary in ways that nonmammalian groups could not. Great apes were beasts of the in-between; captive creatures with a perceived nature rooted in their uncannily human yet distinctly animal appearance and behaviors. In the earliest years when chimpanzees were first displayed at the Zoo and the body of an expired specimen was opened up for scientific scrutiny, there was something distinctly unsettling about the anthropoid, but definitely nonhuman, being lying motionless on the dissecting table. This underlying sense of species proximity in the fluid borderlands between the human and the animal, despite the self-evident animality (or nonhumanness) of the captive creature, was intensified by the percolation of Darwin's insights through both scientific and popular culture and was increasingly evident in the presentation of primate personalities at the Zoo in the 1930s to 1950s through the crafting of celebrity and the staging of tea parties and other apparently hilarious distractions. The power of these animals to mobilize emotions at the heart of the human-animal, object-subject binaries remains today. Its practicality, however, has transformed in line with conservationist agendas. Today's inhabitants of Gorilla Island are clearly positioned as objects of spectacle, scientific scrutiny, and conservation, as well as individuals with almost-human personality traits, explicitly depicted as our closest cousins.

On a smaller scale, the object-subject binary was also specifically configured according to the individual experiences of visitors and staff, some of whom developed highly customized notions of what an animal was and what that animal meant to them. Some visitors perceived animals strictly as objects of study, while others saw them—especially after the middle of the twentieth century—as individuals suffering in prisons (or perhaps, in the case of the pacing bears, in asylums). Some keepers developed relationships with specific animals, as did some visitors, which reflected very deep personal bonds between species, and they did so from the earliest days. Such relationships emerged out of the specific encounters these people had with captive creatures. The recollections and writings of some of the individuals we have encountered in this book's various contexts illustrate the ubiquity of individualization and of the layering of human-animal relationship. Florence Grinfield in her 1894 unofficial guide to the Zoo described a parrot as an object of learning *and* as a pet to whom she gave the name of "Pretty Polly." Some decades later, Alfred the gorilla evoked an array of attitudes that echo the sublime: wonderful and terrifying, tame and irrevocably wild; his skin was branded by the presence of multiple narratives, many of them highly specific. One visitor recalled that Alfred was "fascinating and repulsive, wonderful and terrifying. I loved and hated him but looked forward to seeing him above all the other animals." Some visitors recalled him in the less pleasant contexts of his public defecation and urination, sometimes right onto the visitors themselves.[5]

These histories of relationship orbiting a particular and compellingly singular—yet connected and in many ways representative—captive space reveal much about the multispecies dance of modernity and the apparent *need* for the presence of these layers of relationship. A study of visitors to American zoos and aquaria recently found that a sense of connection with animals, encouraged through the recognition of rich, individual, and autonomous animal lives, positively impacted on visitors' concern about climate change and environments that appear to be remote from their own lives. In turn, this allows conscientious zoos to have impact on the fate of fragile species and ecosystems within and beyond their walls.[6] This much began to become clear from the 1970s at Bristol Zoo, when Oscar the orangutan was adopted by the WWF, now known as the World Wide Fund for Nature, but which was then the World Wildlife Fund. Momentarily objectified to represent the work of the charity and the global plight of threatened species, he even attended a fundraising ball on the charity's behalf. His cute image was at the heart of—even the driver of—the message the WWF wanted to convey.[7] Such layering also allows the Society today to communicate the need to actively involve itself in conservation initiatives, not least in relation to its family group of Western lowland gorillas on Gorilla Island.

Interaction at the heart of these layers also facilitated effective animal husbandry. By imagining animals as both objects of deep zoological knowledge and as subjects with deep emotional lives, keepers came to care for their wards. They could protect eminently predictable bodies and interact with and enrich astonishingly individual personalities. In the process, sentiment grew and provoked intensifications in human commitment to animal lives in captivity. The interlacing of these layers also reflects the troubling ways in which people, past and present, conducted their relationships with animals. Our lives are positively brimming with contradictions reflected in the presence of layers upon layers of meaning and relationship. Animals are everywhere entwined in our lives. They roam among the trees and the grasses, bound and dig among the flowerbeds in our gardens, scuttle and scratch beneath floorboards and in the attic, entering what we often like to think of as human worlds not only as cherished members of an expansive multispecies family but also as terrifying trespassers, loitering eerily in the nooks and crannies. They wander the pages of storybooks and pace the exotic environments rendered captive in the depths of the television screen. Their skins adorn our shivering backs and their flesh is transformed into meat for the hungry: 900 million farm animals are bred for consumption in the UK every year, while 4 million animals are subjected to scientific procedures. These sit alongside some 8.5 million dogs and 7.4 million cats kept as household pets, as well as over five hundred zoos and aquaria attracting throngs of curious visitors.[8] Similar figures abound in the United States, where 8.8 billion chickens, 29 million cattle, and 115 million pigs were slaughtered for food in 2015. At the same time Americans lived with 77.8 million dogs and 85.8 million cats and visited the premises of some 2,500 *licensed* "animal exhibitors."[9]

It is interesting that exploitation sits so centrally in our experience of human-animal encounter, especially in light of the kind of characters so many of us meet in our childhoods: Toad and Ratty in *The Wind in the Willows* (1908); Aslan and the explicitly chimeric Mr. Tumnus in the *Chronicles of Narnia* (1950–56); the beauty of a ravenous caterpillar turned splendid butterfly in *The Very Hungry Caterpillar* (1969); the captivating beasts of *The Gruffalo* (1999); and, more recently, David Walliams's typically chaotic *The Queen's Orangutan* (2015). So many of us seek to kill intruders in our homes; one online retailer features ant-killing products, and its pages are filled with gleeful testimonies to the rapidity and visibility of death the product delivered. There is clearly a separation at work here, a disconnection among the layers of relationship despite the many ways in which they do, in fact, interact with each other.

And so, just like human-animal relationships in the not-so-unnatural world of the zoo, our lives are riddled with layers of contradiction; of science, sentiment, amusement, and commodification entwined. By looking at the zoo as a lens through which to understand transformations in human-animal relationships across the span of modernity, we might conclude that the human-animal relationships therein were, in fact, far from unique or unnatural but were reflective of human-animal interactions in a shared biosphere. Nature and culture, the human and the animal, entwine in all sorts of contexts and are inseparable. Layers of relationship have been at the heart of the common human-animal social fabric since the Zoo's inception and, of course, long before. And they will likely define the human-animal experience into the future. As well as stripping down the natures of such relationships and thus the natures of the beasts themselves, we have also, almost inevitably, turned the mirror of nature-culture upon ourselves. The existence of these layers should tell us that we, too, are fragmentary beasts, dwelling in the world in a state of in-cohesion. We are protean, both as communities of people and as individuals; our minds are characterized by multiplicity and chaos rather than simplicity and consistency.

As I write this, an astonishing number of species are plunging over the cliff edge of extinction, lost to our eyes forever. The WWF's 2016 *Living Planet Report* found that of the ten thousand representative populations of mammals, birds, reptiles, amphibians, and fish studied, numbers have declined by some 58 percent since 1970: "These are the living forms that constitute the fabric of the ecosystems which sustain life on Earth," notes the report, "and the barometer of what we are doing to our own planet, our only home. We ignore their decline at our peril."[10] Indeed, the Bristol Zoological Society's grand plan to grow a megazoo on the outskirts of Bristol recognizes this as one of the great global challenges of our age, and yet it is nevertheless plagued by the layers and contradictions that characterized its zoo and human-animal relationships in modernity. Dedicated to global conservation, research, and education, the new park entered its first phase in 2013 on the Society's Hollywood Tower Estate, a site purchased in 1966.[11] The National Wildlife Conservation Park, as it will come to be known, if all goes to plan, will eventually display an array of endangered species within an ecological framework. State-of-the-art immersive displays will include African plains and semi-aquatic ecosystems featuring wildlife, including piranha, crocodiles, manatees, zebra, rhinoceros, and giraffes, each more captive-and-free than ever before.[12]

Alongside these kinds of exhibits, the park will also display popular (i.e., predominantly large) animals such as tigers and bears that are no longer housed at the Zoo in Clifton. This is another evolution of the zoo, but it retains the layers that defined human-animal relationships in Clifton for nearly two centuries. These creatures are objects of intrinsic value, their genetics holding the key to the survival of species, but they are also of commercial value, for the presence of popular species is vital to the attraction of paying guests. They are spectacular, too, for their animated presence in state-of-the-art, open-air enclosures will both entertain and inform visitors as they learn about the state of the planet while assuaging guilt through the obfuscation of the structures of captivity. As objects of research, these animals' forms and behaviors will be studied, and the results fed into an international community of conservationists and scholars seeking to understand the natural world, not only for the sake of the planet but also for the sake of knowledge itself. Visitors and staff will form attachments; animals will be named and recognized as the individual creatures with rich emotional lives that they are. Nevertheless, they will be fenced in, the human spaces of the new zoo defended from beastly incursions in a series of negotiations among the species that make the place. Layer upon layer upon layer, the new zoo will reflect the human-animal relationships of a multispecies world that is both different from and remarkably similar to the captive worlds that went before it.

I would like to end by recounting a recent family day out. One Saturday in springtime, we took a ride out to the Wild Place Project, the name given to the first phase of this new conservation park. The landscaped, immersive lemur enclosure there allows the visitor to get close to families of ring-tailed, mongoose, and red-bellied lemurs on display. Critically endangered inhabitants of the island of Madagascar, they roam the enclosure space occasionally leaping from tree to tree. Sometimes they do so via the head of a delighted/alarmed visitor. The lemurs are precious emblems of a world on the brink of ecological disaster; they are spectacular objects of fascination; and they are objects of intense scientific study that has already borne fruit in successful captive breeding and that is also leading to the development of ever more effective conservation initiatives. These, like so many creatures who went before them, are entangled in layered relationships with people in their captive world. And my little boy loves and fears them in equal measure. He says he knows what they are thinking and that they are his friends. He cares about them, but he also fears them when they come too close for comfort. Just like Max, then, whose wild place grew up in his bedroom, my boy entered a world within the zoo. He went to "where the wild things are," and there they turned out to be complex beasts indeed; flesh and blood certainly, but they also forged habitats in the mind that make them, and us, much, much more than that.

Notes

INTRODUCTION

1. Davis, *Spectacular Nature,* 8.

2. Bruyninckx, "Majority of Europe's Species."

3. Coates, *Nature.*

4. See Ritvo, "Beasts in the Jungle." For the purposes of this book, the term "wild" will be used—while recognizing its problematic character—to refer to the unfamiliarity of zoo animals and the environments from which they actually—or were perceived to have—emanated.

5. For an introduction to the multifarious ways in which animals inhabit human cultures, see, for instance, Kalof and Resl, *Cultural History of Animals.*

6. Despite the homogenization associated with the umbrella term "animal," this book will nevertheless make use of it for the sake of brevity. It will employ the terms "animal," "creature," "nonhuman animal," and "beast" (despite its rather negative contemporary connotations) interchangeably to denote living things other than plant life and *Homo sapiens.*

7. On the "abyssal rupture," see Aristotle, *De Anima,* II, 3; Derrida, "Animal That Therefore I Am," 399. Also see Haraway, *When Species Meet.*

8. Derrida, "Animal That Therefore I Am," 369–418; Fudge, *Perceiving Animals,* 10.

9. Plumwood, "Being Prey"; Walker, "Animals and the Intimacy." For more on the Ebola epidemic, see Mitman, "Ebola in a Stew," accessed 25 January 2017.

10. Ritvo, *Platypus and the Mermaid,* 131–87.

11. The term "Anthropocene" has recently been coined as a not entirely unproblematic means of suggesting that humans have themselves become a force capable of altering the biosphere. When the Anthropocene began and its ability as a concept to fairly apportion responsibility across time and among cultures have proven to be a source of much conversation: Crutzen and Stoermer, "'Anthropocene,'" 17–18.

12. For a recent introduction to the main concerns of the field of animal studies see Weil, *Thinking Animals.* The field has not, however, been without its detractors. For an example of attitudes toward the expansion and democratization of social history in the 1970s see Phineas, "Household Pets." The subject of the article was not meant to be taken seriously. Written under a pseudonym, it uses the figure of the domesti-

cated animal as a way in which to poke fun at the methods and priorities of historical scholarship at the very margins of human social worlds.

13. The term "speciesism" first appeared in a 1971 essay in which Richard Ryder exposed the cruelties and inequalities inherent in vivisection in Britain and the United States: Ryder, "Experiments on Animals." For expansions upon the concept see Regan, *Case for Animal Rights;* Singer, *Animal Liberation.*

14. Ritvo, "Animal Planet," 204.

15. Lévi-Strauss, *Totemism,* 162.

16. Henninger-Voss, *Animals in Human Histories;* Kalof, *Looking at Animals;* Ritvo, *Animal Estate;* Robbins, *Elephant Slaves.* A particular strand of this work considers the connections between the significations and experiences of animals and those of women in patriarchal systems of being and governing. See Adams and Donovan, *Animals and Women;* Birke, "Intimate Familiarities?"; Emel, "Are You Man Enough"; Merchant, *Death of Nature;* Plumwood, *Feminism and the Mastery;* Wilson, "Sexing the Hyena."

17. Bekoff, *Minding Animals;* Bekoff, *Emotional Lives of Animals.*

18. Bekoff, "Increasing Our Compassion Footprint." A number of scholars have influenced my approach to understanding the historical complexities of human engagements with animals in the past. See Benson, "Animal Writes"; Burt, "Illumination of the Animal Kingdom," 203–4; Fudge, "Left-Handed Blow"; Kean, "Challenges for Historians"; Ritvo, "History and Animal Studies"; Ritvo, "Among Animals"; Specht, "Animal History after Its Triumph"; Wolch and Emel, "Bringing the Animals Back In"; Kean, *Great Cat and Dog Massacre.*

19. Rock, "Never Scared" (emphasis in original).

20. Crosby, *Ecological Imperialism;* V. D. Anderson, *Creatures of Empire.*

21. Haraway, *When Species Meet,* 42; Hearne, *Adam's Task.* Approaches in the social sciences have considered animals as integral parts of expansive social networks: Barad, "Agential Realism"; Callon, "Some Elements of a Sociology"; Latour, *Reassembling the Social;* Whatmore, *Hybrid Geographies.* Also see Woods, "Fantastic Mr. Fox?"

22. A number of works have successfully incorporated both the "semiotic" and the more material approaches to the study of animals past in a holistic approach to human-animal relationships. See, for instance, Howell, *At Home and Astray;* Kete, *Beast in the Boudoir;* Mikhail, *Animal in Ottoman Egypt.* The fields of animal studies and animal history have also spawned a number of important collections: Acampora, *Metamorphoses of the Zoo;* Creager and Jordan, *Human-Animal Boundary;* Nance, *Historical Animal;* Philo and Wilbert, *Animal Spaces, Beastly Places;* Wolch and Emel, *Animal Geographies.*

23. "Hybrid geographies" is a term coined by Sarah Whatmore in her analysis of the influence of the nonhuman in the making of worlds: Whatmore, *Hybrid Geographies.*

24. There are number of works that detail, from a number of vantage points, the histories of zoos across the world. See Baratay and Hardouin-Fugier, *Zoo;* Bender,

Animal Game; Braverman, *Zooland;* Fisher, *Zoos of the World;* Hanson, *Animal Attractions;* Ito, *London Zoo and the Victorians;* Kisling, *Zoo and Aquarium History;* I. J. Miller, *Nature of the Beasts;* Rothfels, *Savages and Beasts.* A number of works "read" the zoo in a more metaphysical sense: Acampora, "Zoos and Eyes"; K. Anderson, "Culture and Nature"; Axelsson and May, "Constructed Landscapes in Zoos"; Lindahl Elliot, "See It, Sense It"; Malamud, *Reading Zoos;* Mullan and Marvin, *Zoo Culture;* Osborne, *Nature, the Exotic, and the Science of French Colonialism;* Uddin, *Zoo Renewal;* Veltre, "Menageries, Metaphors and Meanings." Some works have started to theorize the character of the "drive-thru" safari park. See Flack, "Lions Loose."

25. Hoage, Roskell, and Monsour, "Menageries and Zoos to 1900," 8–18.

26. Bennett and Hodge, *Science and Empire;* Baratay and Hardouin-Fugier, *Zoo,* 113–30; R. W. Jones, "Sight of Creatures"; Ritvo, *Animal Estate,* 205–42.

27. MacKenzie, *Empire of Nature;* Mitchell, *Colonising Egypt,* 6. Also see Logan, *Victorian Parlour.*

28. Bailey, *Leisure and Class in Victorian England;* Bailey, "Politics and Poetics"; H. Cunningham, *Leisure in the Industrial Revolution.*

29. Cowie, *Exhibiting Animals,* ch. 5.

30. There are a number of "institutional" histories of Bristol Zoo: Brown, Ashby and Schwitzer, *Illustrated History of Bristol Zoo Gardens;* Green-Armytage, *Bristol Zoo, 1835–1965;* Warin and Warin, *Portrait of a Zoo.*

31. Meeting 22 July 1835, in General Meetings, July 1835–April 1854, 1. For more on the Bristol Institution for the Advancement of Science, Literature and the Arts, see Barker, *Bristol Museum;* Meller, *Leisure and the Changing City,* 43–45.

32. For a social history of Bristol Zoo and the Bristol, Clifton and West of England Zoological Society see Maddeaux, "'Our Clifton Zoo,'" esp. ch. 2.

33. Branagan, "Stutchbury, Samuel"; Alford, "Wills Family."

34. Loudon, *Suburban Gardener and Villa Companion,* 10.

35. "An Act for extending the Powers of Sale and Exchange contained in the Marriage Settlement of Francis Adams the younger, Esquire; and for other Purposes," 17 July 1837, Anno Primo Victoria Reginae, Cap. 43 (private bill) Bristol Record Office (BRO/43041/2).

36. For an introduction to the history of Clifton see D. Jones, *History of Clifton.*

37. Meeting 22 July 1835, in General Meetings, July 1835–April 1854, 1. On museums, parks, etc. see Lasdun, *English Park;* Macdonald, "Expanding Museum Studies"; Woodson-Boulton, "Victorian Museums and Victorian Society."

38. Neve, "Science in a Commercial City."

39. Pyenson and Pyenson, *Servants of Nature,* 97. *Felix Farley's Bristol Journal,* 16 July 1836; *Bristol Mercury,* 16 July 1836.

40. Keeling, *Where the Lion Trod.* Clinton H. Keeling published prodigiously on the histories of zoological parks in the United Kingdom. Many of his surviving works are held in the archive at the Zoological Society of London or are in private hands.

41. Meeting 6 March 1946, in General Committee and the Executive Council, August 1944–January 1956, 51.

42. *132nd Annual Report of the BCWEZS*, 1 April 1968, 4.

43. Braverman, *Zooland*, 57–58.

44. Meetings 5 July 1836 and 19 July 1836, in Zoology Sub-committee and Fetes Committee, May 1836–September 1850, 7–10.

45. *Felix Farley's Bristol Journal*, 23 July 1836; Hanson, *Animal Attractions*, 44.

46. Acquisition of an Elephant from Balasore, 1912–13; *76th Annual Report of the BCWEZS*, 10 April 1912, 10; *77th Annual Report of the BCWEZS*, 9 April 1913, 9.

47. I. J. Miller, *Nature of the Beasts*.

48. *Manchester Guardian*, 14 September 1842.

49. *Bristol Evening Post*, 23 February 1938, in Newspaper Clippings, c. March 1936–c. July 1961; *Sydney Morning Herald*, 16 June 1950; *103rd Annual Report of the BCWEZS*, 12 April 1939, 6.

50. Meeting 2 June 1869, in General Committee and the Finance Committee, October 1866–September 1880, 96.

51. Meetings 11 April 1838 and 10 April 1839, in General Meetings, July 1835–April 1854, 54, 70–71; Meeting 13 October 1838, in General Committee, April 1836–October 1848, 178.

52. Andrew Kitchener, Head of Birds, Mammals and Taxidermy, National Museums Scotland, email to the author, 4 July 2013.

53. Amato, "The White Elephant in London," 41; Flack, "'Illustrious Stranger,'" 1; *Bristol Evening Post*, 14 March 1975, in Bristol Zoo Newspaper Cuttings 1932–85, Bristol Reference Library.

54. Knocker, *Illustrated Official Guide to the Clifton Zoological Gardens*, 49.

55. *Western Daily Press*, 18 March 1929, in Newspaper Clippings, c. July 1926–c. August 1933; Meeting 4 November 1931, in General Committee, January 1931–July 1944, 29.

56. *129th Annual Report of the BCWEZS*, 7 April 1965, 7; *130th Annual Report of the BCWEZS*, 4 April 1966, 6.

57. Adeney Thomas, "AHR Forum"; Barrow, *Nature's Ghosts;* Chakrabarty "Climate of History."

1. HARVEST AND HERITAGE

1. *Bristol Mercury*, 11 July 1835; *First Annual Report of the BCWEZS*, 1836.

2. MacKenzie, *Empire of Nature*, 47; Ritvo, *Animal Estate*, 47. See also Green, *Tiger*, esp. 68–150; Jackson, *Lion;* Storey, "Big Cats and Imperialism."

3. Meeting 6 July 1838, and Meeting 9 July 1838, in General Committee, April 1836–October 1848, 171–72; Meeting 27 May 1853, in Special and later General Committee, May 1836–April 1856, 142.

4. *Bristol Mercury and Daily Post*, 30 July 1894.

5. *Bristol Mercury*, 25 July 1890; Grinfield, *Our Clifton Zoo*, 5–6, 28.

6. Share certificates, letters to animal dealers and opticians, c. 1893–c. 1894. Incidentally, Barnum and Bailey's "Greatest Show on Earth" announced that it was to close after 146 years in January 2017: "Curtain Comes Down on Ringling Bros and Barnum & Bailey Circus—the Greatest Show on Earth," *Telegraph,* 15 January 2017, accessed 20 January 2017, http://www.telegraph.co.uk/news/2017/01/15/curtain-comes-greatest-show-earth/.

7. Here I draw upon recent work by historians concerned with the development of transnational connectivities and literature relating to imperial science and the mobilities of ideas, specimens, and scientists. See, for instance, Ballantyne, *Orientalism and Race;* Drayton, *Nature's Government;* Grove, *Green Imperialism;* Lambert and Lester, *Colonial Lives.*

8. Meeting 29 November 1961, Meeting 8 April 1963, Meeting 8 July 1963, and Meeting 9 September 1963, in Council Minutes, February 1956–February 1966, 206–7, 249, 260, 265; *Bristol Evening Post,* 2 April 1964, in Newspaper Clippings, c. September 1961–c. April 1965.

9. Meeting 8 July 1968, in Council Minutes, September 1967–December 1983, 21; *Australian Women's Weekly,* 26 August 1970, 42; "Zoo Cubs aka New Arrivals at Bristol Zoo," c. 1968, British Pathé.

10. *Western Daily Press,* 1 July 1971, in Newspaper Clippings, c. March 1963–c. October 1972; Meeting 28 April 1971 and Meeting 30 June 1971 in Council Minutes, September 1967-December 1983, 65, 68.

11. *136th Annual Report of the BCWEZS,* 19 April 1972, 5.

12. Discussion Paper for Animal Committee, 30 October 1985, in Animal Committee, c. 1985–c. 1986; Roychoudhury and Samkhala, "Inbreeding in White Tigers." Also see Kelly, Pearson, and Wright, "Morbidity in Captive White Tigers," 183–88.

13. For further information relating to the genetics of the Asiatic lion, see O'Brien et al., "Evidence for African Origins."

14. Here I employ Collard and Dempsey's term "lively commodities" to denote "live commodities whose capitalist value is derived from their status as living beings": Collard and Dempsey, "Life for Sale?"

15. Warkentin and Fawcett, "Whale and Human Agency."

16. *Catalogue of the Animals.*

17. Lynn, "From Sail to Steam."

18. Headrick, *Tools of Empire,* 151–87; MacKenzie, *Empire of Nature,* 7, 37.

19. Hagenbeck, *Beasts and Men,* 57–71; Rothfels, "Catching Animals." 191–92, 199.

20. Grinfield, *Our Clifton Zoo,* 28.

21. Hartman, *Les singes anthropoïdes et leur organisation comparée à celle de l'homme,* cited in Baratay and Hardouin-Fugier, *Zoo,* 118.

22. Meeting 2 September 1863, in General Committee, June 1861–September 1866, 120; *54th Annual Report,* 9 April 1890, 4.

23. *19th Annual Report of the BCWEZS,* 11 April 1855, 12; *20th Annual Report of*

the BCWEZS, 9 April 1856, 13–14; *Catalogue of the Animals,* 5; *York Herald,* 21 January 1881.

24. Beinart, "Empire, Hunting and Ecological Change," 163–66.

25. Particular regions of the African continent were vital to the industrialization of Europe. Its natural resources—such as palm oil, but also rubber (in West Africa) and, of course, its peoples—were exploited in the name of profit. Over time the extraction of resources led to deforestation, loss of habitats, and a reduction in the continent's biodiversity. For introductory examinations of Africa's environmental histories, see McCann, *Green Land, Brown Land,* part 2; Beinart, "African History and Environmental History." For a recent assessment of the consequences of palm oil plantation and extraction, see "Palm Oil: Cooking the Climate," Greenpeace, 8 November 2007, accessed 29 September 2016, http://www.greenpeace.org/international /en/news/features/palm-oil_cooking-the-climate/.

26. *16th Annual Report of the BCWEZS,* 9 April 1851.

27. Pratt, *Imperial Eyes,* 108.

28. Baratay and Hardouin-Fugier, *Zoo,* 124; Mullan and Marvin, *Zoo Culture,* 138.

29. Imperial networks and webs are themes at the heart of the New Imperial History. For more on imperial scientific networks, see Brockway, *Science and Colonial Expansion;* Laidlaw, *Colonial Connections.*

30. Rothfels, "Catching Animals," 197.

31. Meeting 4 July 1860, in General Committee, June 1861–September 1866, 172; Meetings 6 September 1888, 6 October 1886, and 4 May 1887, in General Committee and the Finance Committee, November 1880–March 1893, 16, 182, 219; Meeting 6 July 1904, in General Committee, April 1893–December 1911, 238; *65th Annual Report of the BCWEZS,* 10 April 1901, 10. For details on the character of some of the networks in which some of these dealers were implicated see Fan, *British Naturalists,* and Rothfels, *Savages and Beasts.*

32. Meeting 15 April 1891, in General Committee and the Finance Committee, November 1880–March 1893, 275.

33. *Glasgow Herald,* 10 April 1872.

34. *1st Annual Report of the BCWEZS,* 1836; Meeting 2 November 1851, in Special and later General Committee, May 1836–April 1856, 77; Meeting 4 October 1865, in General Committee, June 1861–September 1866, 200.

35. On "Animal diplomacy" see Cushing, "Platypus Diplomacy."

36. *The Bristol Presentment: or Bill of Entry,* 35, 23 May 1853, 3566–62, 5 April 1880, 6464; *Port of Bristol HM Customs Bill of Entry,* 1, 7 February 1881, 11–13, 29 June 1893, 1301.

37. *12th Annual Report of the BCWEZS,* 8 April 1847–*134th Annual Report of the BCWEZS ,* 15 April 1970; *Catalogue of the Animals;* Meeting 20 June 1836, in Zoology Sub-committee and Fetes Committee, May 1836–September 1850, 5.

38. "Address," *West of England Journal of Science and Literature,* i–vii; Owen, "On the Extent and Aims."

39. Ritvo, "Order of Nature," 46.

40. *18th Annual Report of the BCWEZS,* 12 April 1854, 7.

41. Knocker, *Illustrated Official Guide,* 1899, 19; Meetings 8 June 1863 and 3 August 1864, in General Committee, June 1861–September 1866, 111, 159; Meeting 7 February 1872, in General Committee and the Finance Committee, October 1866–September 1880, 172; Phillips, *Gardens and Menagerie,* 24.

42. Orwell, *Animal Farm,* 133.

43. Meeting 10 April 1839, in General Meetings, July 1835–April 1854, 63.

44. Mullan and Marvin, *Zoo Culture,* 4.

45. Meeting 2 November 1864, in General Committee, June 1861–September 1866, 167.

46. Flack, "'Illustrious Stranger'"; Meeting 2 May 1860, in General Committee, April 1856–June 1861, 165; Root, "Victorian England's Hippomania."

47. *Bristol Mercury and Daily Post,* 21 March 1892.

48. Spencer, "Pithekos to Pithecanthropus," 13–22.

49. Meetings 12 June 1839 and 1 July 1840, in General Committee, April 1836–October 1848, 212, 239–40; Meeting 14 April 1841, in General Meetings, July 1835–April 1854, 94.

50. *Catalogue of the Animals,* 16–17, 33–34; *Jackson's Oxford Journal,* 18 December 1841.

51. Ritvo, "Calling the Wild," 109–11.

52. "Breed Information Centre: Alaskan Malamute" and "Breed Information Centre: Retriever (Labrador)," Kennel Club, www.thekennelclub.org; "Saint Bernard," *Columbia Encyclopedia,* 6th ed., 2008, http://www.encyclopedia.com/topic/Saint _Bernard_%28dog%29.aspx; *Catalogue of the Animals,* 20; Meeting 5 June 1867, in General Committee and the Finance Committee, October 1866–September 1880, 23.

53. Ritvo, "Beasts in the Jungle."

54. Ritvo, *Animal Estate,* 217.

55. Hanson, *Animal Attractions,* 44.

56. For a delineation of the idea of "charismatic megafauna" see Lorimer, "Nonhuman Charisma."

57. *Bristol Mercury,* 18 September 1858.

58. Baratay and Hardouin-Fugier, *Zoo,* 109, 151.

59. *Bristol Mercury,* 2 July 1881.

60. *Bristol Mercury and Daily Post,* 26 October 1894; *69th Annual Report of the BCWEZS,* 12 April 1905, 12.

61. Baratay and Hardouin-Fugier, *Zoo,* 24; General Committee and the Finance Committee, October 1866–September 1880.

62. *Clifton Chronicle & Directory,* 15 March 1854.

63. *68th Annual Report of the BCWEZS,* 13 April 1904, 4.

64. Meeting 4 June 1890, in General Committee and the Finance Committee, November 1880–March 1893, 254.

65. Mullan and Marvin, *Zoo Culture,* 150.

66. Meetings 7 December 1870 and 1 February 1871, in General Committee and the Finance Committee, October 1866–September 1880, 137, 141.

67. Meeting 9 January 1961 and Meeting 13 February 1961, in Council Minute Book, February 1956–February 1966, 171–73.

68. Meeting 6 February 1884, in General Committee and the Finance Committee, November 1880–March 1893, 74.

69. Meeting 30 October 1839, in General Committee, April 1836–October 1848, 221; Meeting 5 October 1860, in General Committee, April 1856–June 1861, 186.

70. *21st Annual Report of the BCWEZS,* 8 April 1857, 8.

71. Meeting 4 September 1839, in General Meetings, July 1835–April 1854, 217.

72. *43rd Annual Report of the BCWEZS,* 9 April 1879, 10; *50th Annual Report of the BCWEZS,* 14 April 1886, 7.

73. *28th Annual Report of the BCWEZS,* 13 April 1864, 6.

74. *Illustrated Official Guide,* 1894, 25.

75. *Northern Liberator and Champion,* 1 August 1840; *Standard,* 27 July 1840; Altick, *Shows of London,* 362–71.

76. Meeting 3 January 1838, in General Committee, April 1836–October 1848, 105.

77. *85th Annual Report of the BCWEZS,* 13 April 1921, 6.

78. *107th Annual Report of the BCWEZS,* in 14 April 1943, 4.

79. Meeting 3 September 1930, in General Committee, January 1912–December 1930, 344; *Bristol Mercury,* 17 February 1849; du Chaillu, *Explorations and Adventures;* Gott and Weir, *Gorilla,* 44, 97.

80. Blunt, *Ark in the Park,* 137; *Times,* 25 October 1887; Murray, "Lives of the Zoo," 243; *Bristol Mercury,* 21 September 1885.

81. *Milwaukee Journal,* 1 April 1940, 8; *Milwaukee Sentinel,* 2 January 1951, 1; *Pittsburgh Press,* 18 June 1950, 35; *Ludington Daily News,* 2 January 1951, 7.

82. Baratay and Hardouin-Fugier, *Zoo,* 24; BCWEZS Valuation of Livestock, 30 September 1935, in Stock Book: list of stock and valuations with purchase price, 1935–37.

83. Meeting 7 December 1927, in General Committee, January 1912–December 1930, 241.

84. Meetings 9 January and 13 February 1961, in Council Minutes, February 1956–February 1966, 171–73.

85. Ritvo, "Calling the Wild," 107.

86. "Only Six Left Alive," *Zoo,* 52–54.

87. Knocker, *Illustrated Official Guide,* 1899, 15.

88. Vicars-Webb *et al., Guide to the Gardens,* 1926, 45–47.

89. Ibid, 43–47. Incidentally, Britain's last resident golden eagle disappeared from its home in the Lake District and was feared dead in April 2016: "England's Only Resident Golden Eagle 'Feared Dead,'" BBC News, 14 April 2016, accessed 20 January 2017, http://www.bbc.co.uk/news/uk-england-cumbria-36038262.

90. *World Zoos: Bristol,* 23 October 1962, BBC Audio-Visual Archive.

91. Worster, *Nature's Economy*, 341–43.

92. Coates, *Nature*, 125–27.

93. Baratay and Hardouin-Fugier, *Zoo*, 211–24; Burt, *Animals in Film*, 10.

94. Singer, *Animal Liberation*.

95. Export of Penguins from South Georgia for the Bristol Zoo, c. January 1937–c. May 1938, National Archives, Kew (CO78:206/11).

96. K. Anderson, "Culture and Nature," 286.

97. Braverman, *Zooland*, 44.

98. *Times*, 25 July 1968; Meeting 9 June 1975 and 20 September 1976, in Council Minutes, September 1967–December 1983, 111, 123.

99. See, for instance, Meeting 24 September 1985, in Council Meetings, January 1984–December 1994, 31.

100. See Carr and Cohen, "Public Face of Zoos."

101. For a series of conversations around the complex relationships between conservation, commerce, and entertainment in the contemporary zoo, see the essays in Frost, *Zoos and Tourism*.

102. Meeting 15 September 1980, in Council Minutes, September 1967–December 1983, 159; *147th Annual Report of the BCWEZS*, 28 April 1983, 7; Meeting 24 June 1985, in Council Meetings, January 1984–December 1994, 25.

103. Meeting 3 October 1983, in Council Minutes, September 1967–December 1983, 202.

104. Meeting 19 September 1991, in Council Meetings, January 1984–December 1994, 115.

105. *Annual Report of the BCWEZS, 31 December 1992*, 10.

106. Baratay and Hardouin-Fugier, *Zoo*, 272.

107. Braverman, *Zooland*, 174; Warkentin and Fawcett, "Whale and Human Agency," 107.

108. In 1976, San Diego Zoo established the Frozen Zoo® where the ova, semen, blood, cell cultures, and tissue samples of nearly one thousand species are stored in perpetuity. According to San Diego Zoo Institute for Conservation Research, the project, in collaboration with similar initiatives around the world, has the potential to "facilitate advances in our understanding of the biology of endangered species that will directly aid in their conservation and management in the wild": "Frozen Zoo®," San Diego Zoo Research Institute, accessed 20 January 2017, http://institute.sandiegozoo.org/resources/frozen-zoo%C2%AE.

109. "Bristol Zoo Red Panda Gets Mate to Boost Breeding," BBC News, 27 June 2011, accessed 20 January 2017, www.bbc.co.uk/news/uk-england-bristol-13872914; Specimen Report for BRISTOL/11893, Animal Record Keeping System Report, Generated 3 August 2011.

110. Shiva, Lioness 11399, Zoological Information Management System Report, Generated 5 March 2011.

111. John Partridge, email message to author, 13 June 2011.

112. Braverman, *Zooland,* 21.

113. Partridge, email, 13 June 2011.

114. "*This Is Bristol and Bath,*" c. April 1990, in Newspaper Cuttings, 1986–2011; Meeting 26 April 1990, in Council Meetings, January 1984–December 1994, 93; Braverman, *Zooland,* 173; "Marius the Giraffe Killed at Copenhagen Zoo despite Worldwide Protests," *Guardian,* 9 February 2014, accessed 20 January 2017, http://www.theguardian.com/world/2014/feb/09/marius-giraffe-killed-copenhagen-zoo-protests; Ian Parker, "Killing Animals at the Zoo," *New Yorker,* 16 January 2017, accessed 20 January 2017, http://www.newyorker.com/magazine/2017/01/16/killing-animals-at-the-zoo.

115. "*International Union for Conservation of Nature and Natural Resources Policy Statement on Captive Breeding,*" 1 September 1986.

116. "Bristol Zoo Hopes to Save Last Colony of Tree Snail," BBC News, 15 April 2010, accessed 20 January 2017, http://news.bbc.co.uk/1/hi/england/bristol/8621496.stm; *Bristol Evening Post,* 21 February 1998.

117. "Water Vole Population on the Up," BBC News, 12 September 2003, accessed 20 January 2017, http://news.bbc.co.uk/1/hi/england/somerset/3100056.stm.

118. Braverman, *Zooland,* 177.

119. Byers et al., "'One Plan' Approach," 2–5.

120. *Bristol Zoo Gardens Official Guide,* 2007, 37.

121. Baratay and Hardouin-Fugier, *Zoo,* 221.

122. Meeting 6 November 1994, in Education, Research, Conservation and Scientific Committee of Bristol Zoo, June 1986–December 1995.

123. Braverman, *Zooland,* 44.

124. Meeting 26 March 1990, in Council Meetings, January 1984–December 1994, 88.

125. Meeting 15 June 1991, in Council Meetings, January 1984–December 1994, 117.

126. Meeting 15 April 1994, in Council Meetings, January 1984–December 1994, 183–85.

127. Meeting 12 December 1988, in Council Meetings, January 1984–December 1994, 73.

128. "Baby Western Lowland Gorilla Born at Bristol Zoo," BBC News, 28 September 2011, accessed 20 January 2017, http://www.bbc.co.uk/news/uk-england-bristol-15086721; "Rare Lion Cubs Kamran and Ketan Born at Bristol Zoo," ITV News, 23 January 2013, accessed 20 January 2017, *http://www.itv.com/news/2013–01–23/rare-lion-cubs-born-at-bristol-zoo/.*

129. Mullan and Marvin, *Zoo Culture,* 148; "Reindeer Swoop into Bristol Zoo," BBC News, 21 December 2008, accessed 20 January 2017, http://news.bbc.co.uk/1/hi/england/bristol/7794651.stm.

130. Candea, "'I Fell in Love'"; "Compare the Market" (price comparison website), www.comparethemarket.com.

2. THE THEATERS OF ANIMALITY

1. *Catalogue of the Animals.*

2. Barker, *Bristol at Play.*

3. Bieder, *Bear,* 102–15; *Catalogue of the Animals.*

4. *Bristol Zoo Gardens,* 2001.

5. Beardsworth and Bryman, "Wild Animal in Late Modernity," 88–89.

6. On the zoo as a site of mastery of nature (and human Others), see Mullan and Marvin, *Zoo Culture;* Ritvo, *Animal Estate,* 205–88.

7. Phillips, *Gardens and Menagerie,* 8, 14; Knocker, *Illustrated Official Guide,* 1903, 19.

8. Meeting 8 April 1840, in General Meetings, July 1835–April 1854, 75.

9. See *Catalogue of the Animals;* Knocker, *Illustrated Official Guide,* 1899.

10. Melville, "Visit of St Mary Redcliff National Schools"; Grinfield, *Our Clifton Zoo,* 15; W. E. Cecil to Curator, 1 May 1911, in Subscriptions, complimentary entrance tickets and general enquiries, c. 1911.

11. Driver and Gilbert, *Imperial Cities,* 1.

12. Baratay and Hardouin-Fugier, *Zoo,* 120; Mitman, "When Nature *Is* the Zoo," 118–21. There is an ongoing debate around the nature of the impact of the British Empire on the domestic stage, though it is beyond the scope of this work. See Burton, *At the Heart;* MacKenzie, *Imperialism and Popular Culture;* Porter, *Absent-Minded Imperialists;* Thompson, *Empire Strikes Back?*

13. K. Anderson, "Culture and Nature," 282; Braverman, *Zooland,* 30.

14. *Catalogue of the Animals,* 5, 7, 9, 12, 15, 31; Phillips, *Gardens and Menagerie,* 8–9, 12, 15, 18, 20, 48.

15. Everest, "'Under the Skin,'" 80; *Bristol Zoo Book,* 1948, 11.

16. *116th Annual Report of the BCWEZS,* 10 March 1952, 5.

17. Wylie, *Elephant,* 118–19.

18. *Bristol Zoo Book,* 1954, 11; Hartley, *Bristol Zoo Book,* c. 1973, 17.

19. Braverman, *Zooland,* 71.

20. This marriage of sight and knowledge in the context of the zoo echoes the dynamic underpinning Jeremy Bentham's Panopticon. See Foucault, *Discipline and Punish.*

21. *Bristol Museum Annual Reports,* 1891–1907; Plumb, "'In Fact,'" 29.

22. Meeting 26 May 1853, in Special and later General Committee, May 1836–April 1856, 141–42.

23. Meeting 5 June 1867, in General Committee and the Finance Committee, October 1866–September 1880, 23; Grinfield, *Our Clifton Zoo,* 34; Meeting 8 June 1857, in General Committee, April 1856–June 1861, 35–36. Reference to museum-style specimens and a designated museum space in the Gardens had disappeared from the Society's official literature by 1925.

24. Braverman, *Zooland,* 8; Carruthers, "Enlightenment Science of Society," 239.

25. Foucault, *Discipline and Punish,* esp. 249–51; Mullan and Marvin, *Zoo Culture,* 42–43.

26. "Barless Cage," c. 1950, British Pathé.

27. *118th Annual Report of the BCWEZS,* 18 March 1954, 4.

28. *Times,* 4 March 1954, in Newspaper Clippings, c. March 1936–c. July 1961; Knight, *Animals after Dark,* 19–20.

29. Poliquin, *Breathless Zoo;* Quinn, *Windows on Nature.*

30. Tuan, *Space and Place,* 83.

31. Wang, "Rethinking Authenticity."

32. *Catalogue of the Animals,* 14; Knocker, *Illustrated Official Guide,* 1903, 13; Vicars-Webb *et al., Guide to the Gardens,* 1928, 16.

33. *Times,* 4 March 1954, in Newspaper Clippings Folder, c. March 1936–c. July 1961.

34. *Catalogue of the Animals,* 5, 9.

35. Grinfield, *Our Clifton Zoo,* 23; Melville, "Visit of St Mary Redcliff National Schools"; *103rd Annual Report of the BCWEZS,* 12 April 1939, 7.

36. Bousé, *Wildlife Films,* 44.

37. Knight, "Feeding Mr. Monkey," 231–53.

38. Grinfield, *Our Clifton Zoo,* 12–13.

39. Knocker, *Illustrated Official Guide,* 1899, 21.

40. What follows has been much influenced by a range of scholars undertaking important work in the way of uncovering multisensory modes of being in the world, including the following: Classen, *Book of Touch;* Coates, "Strange Stillness of the Past"; Corbin, *Foul and the Fragrant;* Hoffer, *Sensory Worlds;* Picker, *Victorian Soundscapes;* Smith, *Hearing History.* On taste in the zoo context, see Lever, *They Dined on Eland.*

41. Ritvo, *Animal Estate,* 218.

42. *Catalogue of the Animals,* 34; *Bristol Evening World,* 18 August 1937, in Newspaper Clippings, c. September 1933–c. August 1937; Vanderplank, *Official Guide to the Gardens;* Mike Colbourne, interview, Bristol, 16 March 2011.

43. Grinfield, *Our Clifton Zoo,* 23, 30–31; *Atalanta,* 1 February 1889; Melville, "Visit of St Mary Redcliff National Schools."

44. Grinfield, *Our Clifton Zoo,* 15, 27; *Royal Cornwall Gazette Falmouth Packet, Cornish Weekly News, & General Advertiser,* 5 January 1899, 2; Hahn, *Tower Menagerie,* 173–74.

45. Meeting 7 April 1860, in General Committee, April 1856–June 1861, 154; Meeting 1 January 1862, in General Committee, June 1861–September 1866, 27; Meeting 4 March 1874 and 7 January 1880, in General Committee and the Finance Committee, October 1866–September 1880, 242, 404.

46. Baratay and Hardouin-Fugier, *Zoo,* 185; Walvin, *Child's World,* 182–85; Don Packham, interview, 29 January 2013.

47. Schafer, "Soundscapes and Earwitnesses," 6–9.

48. *Catalogue of the Animals,* 16, 34; *Bristol Zoo Book,* 1957, 14.

49. Grinfield, *Our Clifton Zoo,* 24; Webb, *Webb's New Guide to Bristol,* 1.

50. Rothfels, "Touching Animals," 41–48.

51. Knocker, *Illustrated Official Guide,* 1899, 51.

52. Edwards, *London Zoo from Old Photographs,* 65; *79th Annual Report of the BCWEZS,* 14 April 1915, 10; *Bristol Evening World,* 22 February 1938, in Newspaper Clippings, c. March 1936–c. July 1961.

53. See, for instance, Meeting 6 February 1929, in General Committee, January 1912–December 1930, 275; Meeting 20 November 1935, in General Committee, January 1931–July 1944, 160; Bristol Zoo Gardens Historical Footage, c. 1930–c. 1995.

54. Baratay and Hardouin-Fugier, *Zoo,* 150; Braverman, *Zooland,* 72.

55. Meeting 13 June 1955, in General Committee and Executive Council, August 1944–January 1956, 332; Melville, "Visit of St Mary Redcliff National Schools"; Peter Floyd, interview, 21 November 2012; Nick Hood, interview, 25 February 2013.

56. Meeting 13 June 1955, in General Committee and Executive Council, August 1944–January 1956, 332.

57. Hediger, *Man and Animal,* 216; Bryan Turner to the Society, 25 September 1978, in Complimentary letters c. 1974–c. 1978; Mr. Culling and Family to the Society, 4 October 1978, in Complimentary letters c. 1974–c. 1978; Grinfield, *Our Clifton Zoo,* 9.

58. Baratay and Hardouin-Fugier, *Zoo,* 234–70.

59. Ritvo, *Animal Estate,* 125–66; Turner, *Reckoning with the Beast,* 81–95; *Bristol Mercury and Daily Post,* 1 February 1883, 6 October 1885, 10 October 1885, 18 January 1890. Also see French, *Antivivisection;* Kean, *Animal Rights;* Rupke, *Vivisection in Historical Perspective.* For insights on the Victorian relationship with meat, see Gregory, *Of Victorians and Vegetarians.*

60. Meetings 6 and 7 June 1906, in General Committee, April 1893–December 1911, 265–72.

61. Rothfels, *Savages and Beasts;* Rothfels, "Immersed with Animals," 216; Rothfels, "How the Caged Bird Sings," 103–12.

62. *Bristol Zoo Book,* 1948, 7; *95th Annual Report of the BCWEZS,* 8 April 1931, 8.

63. *Times,* 14 August 1969, in Newspaper Clippings, c. March 1963–c. October 1972; Denise Aldred to the Director, 4 July 1978; Complimentary letters from visitors praising Bristol Zoo, c. 1974–c. 1978. Also see Ellenberger, "Mental Hospital."

64. *135th Annual Report of the BCWEZS,* 19 April 1971, 5; Cat complex 1977 plans, drawings, relevant correspondence, etc., c. 1977; *Western Daily Press,* 23 September 1983, in Newspaper Clippings, c. August 1976–c. November 1986.

65. *Western Daily Press,* 14 February 1991, in Newspaper Cuttings, 1986–2011.

66. *Bristol Zoo Gardens,* 2001; Hancocks, *Different Nature,* 177.

67. For more on the Society's conceptualization of "immersive" displays, see *An-*

nual Review, 2006, 8; Mata Atlantica (Zona Brasil) Dutton Merrifield campaign brief, 1 February 2002.

68. Interpretation: General Points, March 1996, in Twilight World: sound system, suppliers, interpretation and services, c. 1996–c. 1999; Outline proposal for set design for "the dark house," 1995, and Architect's Design, in Twilight World: set designers, publicity and launch, planning, consents, trees and planting, c. 1996–c. 1999.

69. "Gorillas in your Midst: Zoo Unveils £1 Million Glass Enclosure so Visitors Can Get Up Close and Personal with the Primates," *Daily Mail Online,* 23 August 2013, accessed 20 January 2017, http://www.dailymail.co.uk/news/article-2400721 /Gorillas-midst-Zoo-unveils-1million-glass-enclosure-visitors-close-personal -primates.html#ixzz4WIxwYL2c.

70. *Bristol Zoo Matters,* 2000, 8; "Live GorillaCam, Bristol Zoo, England," You Tube, 2 October 2008, accessed 20 January 2017, https://www.youtube.com /watch?v=gyxTL6XJAC8.

71. Baratay and Hardouin-Fugier, *Zoo,* 185; Vanderplank, *Official Guide to the Gardens,* 8; *115th Annual Report of the BCWEZS,* 18 January 1951, 5; *Bristol Zoo Book,* 1957, inside cover.

72. *132nd Annual Report of the BCWEZS,* 1968, 8.

73. John Partridge, email message to the author, 4 March 2013.

74. *Bristol Zoo Book,* 1964, inside front cover; *London Zoo Guide,* 1958, insert added, 1960.

75. Simon Garrett, email message to the author, 21 November 2012.

3. THE GREAT EXPERIMENT

1. See, for instance, *Barrier Miner,* 2 February 1951, 12; *Sydney Morning Herald,* 28 December 1954, 12.

2. Hill, *Sunshine and the Open Air.*

3. "Beryl Corner," *Telegraph,* 20 March 2007, accessed 22 January 2017, http:// www.telegraph.co.uk/news/obituaries/1546038/Beryl-Corner.html; Crossley-Evans, "Obituary—Dr Beryl Corner, 1910–2007," University of Bristol News, 19 March 2007, accessed 22 January 2017, http://www.bris.ac.uk/news/2007/5352.html.

4. "Blind Gorilla Sees Again," 12 April 2002, accessed 22 January 2017, http:// news.bbc.co.uk/1/hi/sci/tech/1921738.stm; *Annual Review,* 2006; *Bristol Zoo Matters,* Spring 2007, 5. "Western Lowland Gorilla Born at Bristol Zoo," Newsround, BBC, 28 September 2011, accessed 22 January 2017, http://www.bbc.co.uk /newsround/15092928; "Baby Gorilla Born at Bristol Zoo Gardens after Rare Caesarean Section," Bristol Zoo Gardens, 23 February 2016, accessed 22 January 2017, *http://www.bristolzoo.org.uk/latest-zoo-news/Baby-gorilla-born-by-rare-caesarean.*

5. Meeting 22 July 1835, in General Meetings, July 1835–April 1854, 1.

6. Meeting 18 March 1954, in General Committee and Executive Council, August 1944–January 1956, 297.

7. On the cultural status of animal specimens see, for instance, Birke, "Animal Bodies," 3–4, 156–78; Crist, *Images of Animals,* 88.

8. *Catalogue of the Animals,* 4.

9. See, for instance, Phillips, *Gardens and Menagerie,* 8–9, 12, 15, 18, 20, 48.

10. *Manchester Guardian,* 7 June 1837, 2; *Manchester Times and Gazette,* 9 June 1838, 3; Meeting 5 July, in Finance and Zoology Sub-committee, 1836–1843; Meetings 5 July 1836, 21 July 1836, 2 August 1836, and 23 November 1836, in Zoology Sub-committee and Fetes Committee, May 1836–September 1850, 7, 11, 13, 15.

11. Gardiner, "'Dangerous' Women of Animal Welfare," 467; Hardy, "Professional Advantage and Public Health," 1–23.

12. Woods and Matthews, "'Little, If at All,," 29–54. For a brief history of the development of zoo veterinary training and practice, see Fowler, "Historical Perspective," 326–30; Gardiner, "'Dangerous' Women," 472.

13. Bennett, "Vets," 132; Clarke, "Spooner, Charles (1806–1871)."

14. Meeting 13 July 1844, in General Committee, April 1836–October 1848, 324. Sadly, no further information survives about Webb's career.

15. Bennett, "Vets," 133.

16. Meeting 7 November 1864, in General Committee, 5 June 1861–September 1866, 167; Meyer, "New Animal Hospital," 47–50.

17. Meeting 2 February 1870 and 1 April 1874, in General Committee and the Finance Committee, October 1866–September 1880, 115, 246; Braverman, *Zooland,* 8.

18. *16th Annual Report of the BCWEZS,* 1851; *25th Annual Report of the BCWEZS,* 10 April 1861.

19. Grinfield, *Our Clifton Zoo,* 23; For more on death in the zoo, see Benbow, "Death and Dying."

20. Jupp and Gittings, *Death in England,* 240.

21. Asma, *Stuffed Animals,* 27.

22. *12th Annual Report of the BCWEZS,* 1847.

23. Meeting 5 June 1861, in General Committee, June 1861–September 1866, 1.

24. Baratay and Hardouin-Fugier, *Zoo,* 134.

25. *Standard,* 16 December 1841.

26. *1st Annual Report of the BCWEZS,* 1836; Baratay and Hardouin-Fugier, *Zoo,* 99; Barrington-Johnson, *Zoo,* 24–25; Ritvo, *Animal Estate,* 235.

27. *1st Annual Report of the BCWEZS,* 1836. The various European zoological societies of the first half of the nineteenth century likewise privileged the provision of such enclosure conditions: Baratay and Hardouin-Fugier, *Zoo,* 99. See also Barrington-Johnson, *Zoo,* 24–25.

28. Meeting 21 July 1836, in Special and later General Committee, May 1836-April 1856, 11.

29. *17th Annual Report of the BCWEZS,* 1853; *Illustrated Official Guide to the Gardens,* 21.

30. Meeting 4 September 1852, in Special and later General Committee, May 1836–April 1856, 117; Meeting 7 August 1872, in General Committee and the Finance Committee, October 1866–September 1880), 189; Bartlett, *Wild Animals in Captivity,* 55.

31. For an introduction to environmental enrichment, see Mellen, Shepherdson, and Hutchins, "Epilogue," 334.

32. Meeting 26 July 1836, in Zoology Sub-committee and Fetes Committee, May 1836–September 1850, 12.

33. Hediger, *Man and Animal,* 276; Hediger, *Wild Animals in Captivity,* 37. Also see Ritvo, *Animal Estate,* 234; Hancocks, *Different Nature,* 79.

34. *18th Annual Report of the BCWEZS,* 12 April 1854, 7.

35. Bartlett, *Wild Animals in Captivity,* 356; Meetings 27 May 1836, 23 November 1836, and 2 December 1836, in Zoology Sub-committee and Fetes Committee, May 1836–September 1850, 2, 16; *Catalogue of the Animals,* 33–34.

36. Meeting 5 January 1870, in General Committee and the Finance Committee, October 1866–September 1880, 1–10.

37. Bostock, *Zoos and Animal Rights,* 81; Hediger, *Wild Animals in Captivity,* 37.

38. Buckland, *Curiosities of Natural History,* 219.

39. *30th Annual Report of the BCWEZS,* 11 April 1866, 6; *33rd Annual Report of the BCWEZS,* 14 April 1869, 6; *34th Annual Report of the BCWEZS,* 13 April 1870, 6.

40. Meeting 25 April 1944, in General Committee, January 1931–July 1944, 395–96; Richard Hewer, interview, 5 July 2013.

41. Meeting 29 June 1836, in General Committee, April 1836–October 1848, 26.

42. Hediger, *Wild Animals in Captivity,* 1.

43. Meeting 7 November 1928, in General Committee, January 1912–December 1930, 267; Meeting 21 December 1932 in General Committee, January 1931–July 1944, 70.

44. Halloran, "Bibliography," 1955. For more information on zoo-based veterinary literature, see Fiennes, "Zoo Medicine," 674–76.

45. Allen, *Life Science,* 9.

46. Davies, "Networks of Nature," 98–101; Lorenz, "Innate Forms of Potential Experience,"; Fericean *et al.,* "History and Development of Ethology."

47. See, for instance, Fossey, *Gorillas in the Mist;* Schaller, *Serengeti Lion.*

48. *Bristol Zoo Book,* 1948, 48; Hartley, *Bristol Zoo Book,* 1973.

49. According to Andrew Cunningham, the insights of the "Bacteriological Revolution" had in fact taken decades to trickle down through the strata of society. This explains why approaches to zoo animals informed by Germ Theory took so long to take root: Cunningham, "Transforming Plague." Also see Worboys, "Was There a 'Bacteriological Revolution'?"

50. 19 March 1953, in Animal Reports, 1923–69; 13 October 1970, in Animal Reports, 1970–93. The Veterinary School at Langford was founded in 1949, after the second Loveday Report (1944) recommended the foundation of two new veterinary schools in the UK: Webster, "University of Bristol."

51. Meeting 16 May 1934, in General Committee, January 1931–July 1944, 116; Vaccination Schedule, 30 July 1975, in Correspondence Files, Vets and professors of livestock, c. September 1970–c. December 1982.

52. Fiennes, "Zoo Medicine," 674–76.

53. Port Mortem Reports, 1976–1988.

54. Fiennes, "Zoo Medicine," 674–76.

55. Report on monkeys 11 April 1928, in Post Mortem Reports, 1928.

56. Ansell, *Pasteurella Pseudotuberculosis.*

57. Pearson and Wright, "Some Observations on the Rearing," 134–36.

58. *Bristol Museum and Art Gallery, Report of the Museum and Art Gallery, 7.*

59. *Times,* 23 August 1968, in Newspaper Clippings, c. March 1963–c. October 1972.

60. Meeting 21 July 1836, in Special and later General Committee, May 1836-April 1856, 11; Meeting 3 August 1881, in General Committee and the Finance Committee, November 1880-March 1893, 16.

61. Meeting 30 May 1853, in Special and later General Committee, May 1836–April 1856, 143; P. Chalmers Mitchell, Secretary of the Zoological Society of London, to The Superintendent, 11 March 1911, in Correspondence, Other UK zoos, c. 1911.

62. Hyson, "Urban Jungles," 449.

63. *118th Annual Report of the BCWEZS,* 18 March 1954, 4.

64. Hediger, *Wild Animals in Captivity,* 104; *132nd Annual Report of the BCWEZS,,* 1 April 1968, 4; *147th Annual Report of the BCWEZS,* 28 April 1983, 6.

65. Meeting 12 August 1963, in Council Minutes, February 1956–February 1966, 260; Meeting 12 June 1967, in Council Minutes, March 1966–August 1967, 45; Meeting 28 February 1968, in Animal Sub-committee, September 1967–December 1977, 10; Meeting 11 May 1970, in Council Minutes, September 1967–December 1983, 66.

66. Sayer, "Animal Machines," 480.

67. Boyd-Orr, *Food, Health, and Income;* Vanderplank, *Official Guide to the Gardens,* 9; *133rd Annual Report of the BCWEZS,* 21 April 1969, 4.

68. *Bristol Evening Post,* 7 November 1944, in Newspaper Clippings, c. March 1936–c. July 1961; *99th Annual Report of the BCWEZS,* 10 April 1935, 7; *117th Annual Report,* 12 March 1953, 4; *123rd Annual Report of the BCWEZS,* 17 March 1959, 4; *131st Annual Report of the BCWEZS,* 6 April 1967, 6; *133rd Annual Report of the BCWEZS,* 21 April 1969, 7. *Bristol Evening Post,* 1 May 1971, in Newspaper Clippings, c. March 1963–c. October 1972; Mike Colbourne, interview, 16 March 2011.

69. *Bristol Zoo Gardens,* 1987, esp. 31; *Bristol Zoo Gardens,* 2001, esp. 12, 18–19.

70. Meeting 10 November 1980, in Council Minutes, September 1967–December 1983, 161; *Annual Report of the BCWEZS,* 2000, 10; Michelle Barrow, head veterinarian, Bristol Zoo Gardens, email to the author, 8 June 2013.

71. *Annual Report of the BCWEZS,* 1990, 7; *Annual Review,* 2000, 10.

72. John Partridge, curator of animals, Bristol Zoo Gardens, email to the author, 15 December 2012.

73. Burns, "Examining the Hormonal and Behavioural Effects"; Tidabock, "Captive Breeding Management."

74. Colby, Mattacks and Pond, "Gross Anatomy," 267–75; correspondence from Caroline M. Pond to Mr. G. Greed, 12 March 1989, in Polar bears letters, zoo check

newsletter, letters from other zoos, c. 1987–c. 1999; Caroline M. Pond, The Open University, email to the author, 1 October 2012.

75. Kitchener and Groves, "New Insights," 533–42; Brassey et al., "Finite Element Modelling," 1–11.

76. See, for instance, Haraway, *Primate Visions,* 122–24.

77. Broom, "History of Animal Welfare Science," 132.

78. Gordon Bland, interview, Bristol, 16 November 2011.

79. John Partridge, curator of animals, Bristol Zoo Gardens, email to the author, 14 December 2011.

80. "Hungry Pandas Need Gardeners' Aid," BBC News, 11 May 2004, accessed 22 January 2017, news.bbc.co.uk/2/hi/uk_news/england/bristol/somerset/3709207.stm.

81. "Endangered Frogs Get 'Love Shack' at Bristol Zoo," BBC News, 11 February 2010, accessed 22 January 2017, http://news.bbc.co.uk/1/hi/england/bristol/8511225.stm.

82. *Annual Review,* 2006, 9; *Bristol Zoo Matters,* Spring 2004, 7; *Bristol Zoo Matters,* Winter 2004, 4.

4. MINDING THE HUMAN-ANIMAL BORDERLANDS

1. *136th Annual Report of the BCWEZS,* 19 April 1972, 6; 29 August 1971, 3 September 1971, 27 September 1971, and 13 November 1971, all in Animal Report, 1971; 2 September 1972, in Animal Report, 1972; *Western Daily Press,* 18 May 1971, in Newspaper Clippings, c. March 1963–c. October 1972; Mike Colbourne, interview, Bristol, 16 March 2011.

2. Bekoff, "Increasing Our Compassion Footprint," 772.

3. Kean, "Challenges for Historians," S67.

4. *Bristol Evening Post,* 21 December 1973, in Newspaper Clippings, c. March 1973–c. April 1974.

5. *Sun,* 4 September 1971, in Newspaper Clippings, c. March 1963–c. October 1972.

6. Sorabji, *Animal Minds and Human Morals.*

7. Baratay and Hardouin-Fugier, *Zoo,* 216–19.

8. Morris, *Naked Ape,* 226–29.

9. Desmond, *Staging Tourism,* 166; Burghardt and Herzog, "Animals, Evolution, and Ethics," 129–51; also see Nash, "Desperately Seeking Charisma"; Stutesman, *Snake,* 137.

10. *Programme for Bristol Hospitals Summer Carnival,* 27.

11. W. M. Jones, *Jarge Balsh at Bristol Zoo,* 109.

12. *Bristol Evening Post,* 1 May 1971, in Newspaper Clippings, c. March 1963–c. October 1972.

13. See Gen. 1:26 and 1:27: "Then God said, 'Let us make humankind in our image, according to our likeness; and let them have dominion over the fish of the sea, and over the birds of the air, and over the cattle, and over all the wild animals of the earth, and over every creeping thing that creeps upon the earth.' So God created humankind in his image."

14. Ritvo, *Animal Estate*, 30.

15. *Catalogue of the Animals,* 4–5, 8, 10, 16, 28, 34; Phillips, *Gardens and Menagerie,* 14–16.

16. *Catalogue of the Animals,* 6–7; Phillips, *Gardens and Menagerie,* 11, 34–49.

17. *Catalogue of the Animals,* 14.

18. *Manchester Guardian,* 14 September 1842.

19. Hahn, *Tower Menagerie,* 195–96; Knight, introduction to *Animals in Person,* 2.

20. Meeting 20 April 1853, in Special and later General Committee, May 1836–April 1856, 138; Meeting 3 September 1879, in General Committee and the Finance Committee, October 1866–September 1880, 396; Meeting 3 January 1894, 6 June 1894, and 3 October 1894, in General Committee, April 1893–December 1911, 32, 45, 51.

21. Fudge, *Animal,* 32; Sanders, "Actions Speak Louder than Words," 410.

22. Braverman, *Zooland,* 20–21.

23. *Jackson's Oxford Journal,* 18 December 1841; Meeting 18 August 1841, in General Committee, April 1836–October 1848, 258–59; Meeting 5 January 1870, in General Committee, April 1856–June 1861, 114.

24. Meeting 17 April 1839, in General Committee, April 1836–October 1848, 204; Meeting 3 September 1884 and Meeting 5 November 1884, in General Committee and the Finance Committee, November 1880–March 1893, 104–8.

25. Lloyd Morgan, "On the Study of Animal Intelligence," 181; *Clifton Chronicle & Directory: A Weekly Register of Local Intelligence, Arrivals, Removals, Departures, Amusements, Etc.* 10 June 1885.

26. Fudge, *Perceiving Animals,* 17.

27. *Bristol Mercury and Daily Post,* 31 May 1878; *Birmingham Daily Post,* 8 June 1878; Lopez, *Of Wolves and Men,* 320.

28. Haraway, *Primate Visions,* 1.

29. Aristotle, *History of Animals,* Book II, Parts 8–9; von Linné, *Systema naturæ;* Tyson, *Orang-outang, sive Homo sylvestris.*

30. Darwin, *Origin of Species;* Darwin, *Descent of Man;* Darwin, *Expression of the Emotions;* Huxley, *Man's Place in Nature;* Owen, *Memoir on the Gorilla.*

31. *Catalogue of the Animals,* 32–33; Cicero, *De Natura Deorum Academica,* 95–96; Xenophon, *Memorabilia of Socrates,* 1.4.II.

32. *Catalogue of the Animals,* 5.

33. *Sheffield & Rotherham Independent,* 25 December 1841.

34. Brunner, *Ocean at Home,* 8, 46, 85; Sax, *Mythical Zoo,* 211; Wadewitz, "Are Fish Wildlife?," 423–27.

35. Phillips, *Gardens and Menagerie,* 48.

36. Brunner, *Ocean at Home,* 11, 119.

37. Vanderplank, *Official Guide to the Gardens,* 20–21; Vicars-Webb *et al., Guide to the Gardens,* 1926, 19.

38. Phillips, *Gardens and Menagerie,* 39.

39. Grinfield, *Our Clifton Zoo,* 15, 22; Bantock, *Last Smyths of Ashton Court,* 195.

40. *Catalogue of the Animals,* 14; Knocker, *Illustrated Official Guide,* 1899.

41. M. Anderson and Henderson, "Pernicious Portrayals," 297–314; Cartmill, *View to a Death,* 138; Hyson, "Urban Jungles," 368, 401; Mitman, *Reel Nature,* 137–54. On pet keeping, see, for instance, Grier, *Pets in America;* Ritvo, "Emergence of Modern Pet-Keeping."

42. Vanderplank, *Official Guide to the Gardens,* 9–11, 45.

43. Vanderplank, *Official Guide to the Gardens,* 15, 17, 20, 23, 33.

44. See, for instance, Knocker, *Illustrated Official Guide,* 1899, 21; Meeting 6 October 1920, in General Committee, January 1912–December 1930, 108; Meeting 17 September 1941, in General Committee, January 1931–July 1944, 337.

45. *Illustrated Official Guide to the Gardens,* 21.

46. *Bristol Evening World,* 11 March 1938, in Newspaper Clippings, c. March 1936–c. July 1961.

47. Knocker, *Illustrated Official Guide,* 1903, 36; *Illustrated Official Guide,* 1904, 33–34; *Illustrated Official Guide to the Gardens,* c. 1924–25, 23.

48. Haraway, *Primate Visions,* 126. Also see Vickery, "Golden Age to Separate Spheres?"

49. Vicars-Webb *et al., Guide to the Gardens,* 1928, 11; *Bristol Evening Post,* 19 February 1935, in Newspaper Clippings, c. September 1933–c. August 1937; *Bristol Evening Post,* c. 1936, in Newspaper Clippings, c. September 1933–c. August 1937.

50. *Daily Sketch,* 10 February 1945, in Newspaper Clippings, c. March 1936–c. July 1961; *Daily Express,* 10 May 1934, in Newspaper Clippings, c. September 1933–c. August 1937.

51. *Bristol Times and Mirror,* 9 December 1927, in Newspaper Clippings , c. July 1926–c. August 1933; *Bristol Evening World,* 24 November 1938, in Newspaper Clippings, c. March 1936–c. July 1961.

52. *Western Daily Press,* 15 October 1935, in Newspaper Clippings, c. September 1933–c. August 1937.

53. Marsh, "Rise of Celebrity Culture."

54. Chambers, *Jumbo;* Flack, "'Illustrious Stranger'"; Jolly, *Jumbo;* Ritvo, *Animal Estate,* 228.

55. Gregory and Raven, *In Quest of Gorillas,* 187.

56. *Bristol Times and Mirror,* 9 April 1931, in Newspaper Clippings, c. July 1926–c. August 1933; *Bristol Evening Post,* 13 January 1934, in Newspaper Clippings, c. September 1933–c. August 1937.

57. Gott and Weir, *Gorilla,* 114–17; Paddon, "Biological Objects," 141–46.

58. Less, Kuhar, and Lukas, "Assessing the Prevalence and Characteristics."

59. *Bristol Evening World,* 5 September 1935, and *Bristol Evening Post,* 5 September 1935, in Newspaper Clippings, c. September 1933–c. August 1937; *Bristol Evening Post,* 13 May 1938, in Newspaper Clippings, c. March 1936–c. July 1961; Vanderplank, *Official Guide to the Gardens,* 9.

60. Hornaday, "Gorillas, Past and Present."

61. *Bristol: For U.S. Armed Forces in U.K; Ellesmere Guardian,* 17 September 1940; *Spokane Daily Chronicle,* 1 August 1940, 2; *Hartford Courant,* 6 November 1940'; *Free Lance-Star,* 31 July 1940, 9; *Pittsburgh Press,* 15 September 1940; *Bristol Evening Post,* 18 March 1946, in Newspaper Clippings, c. March 1936–c. July 1961.

62. *Bristol Evening Post,* c. March 1946, in Newspaper Clippings, c. March 1936–c. July 1961.

63. *Weekend Guardian,* 23–24 June 1990, in Alfred file, c. 1930–c. 2010, BMAG.

64. Paddon, "Biological Objects," 141–46; Tipper, "'Dog Who I Know Quite Well'"; Mr. John Davies to Mr. Ray Barrett, City Museum, 24 May 1999; Memories of Alfred Mrs. M. Claridge, 24 June 1993, Memories of Alfred Mrs. J. V. Kemsley, 15 June 1993, Memories of Alfred, Cynthia Pearson, 20 July 1993, in Alfred file, c. 1930–c. 2010, BMAG.

65. *Western Daily Press and Bristol Mirror,* 10 March 1948; *Bristol Evening Post,* 9 March 1948, both in Newspaper Clippings, c. March 1936–c. July 1961.

66. *Bristol Zoo Book,* 1954, 15. For work on the creation of spaces of memory for animal others, see Kean, "Moment of Greyfriars Bobby," and Kean, "Human and Animal Space."

67. *Bristol Evening Post,* 9 May 1936, *Bristol Evening World,* 9 May 1936, *Bristol Evening Post,* 7 May 1935, and *Bristol Evening World,* 9 May 1935, all in Newspaper Clippings, c. September 1933–c. August 1937.

68. *Bristol Evening Post,* c. 1940, in Newspaper Clippings, c. March 1936–c. July 1961. For Jumbo, see Chambers, *Jumbo;* Jolly, *Jumbo.*

69. Adoption Scheme receipts, 1940–41.

70. Young, *Darwin's Metaphor,* 15.

71. Meetings 21 September 1938, 17 September 1941, and 17 June 1942, in General Committee, January 1931–July 1944, 244, 337, 353; Meeting 11 February 1952 and 14 March 1955, in General Committee and Executive Council, August 1944–January 1956, 228, 327.

72. *Bristol Times and Mirror,* 30 August 1927 and 16 August 1927, and *Clifton Chronicle,* 18 August 1927, in Newspaper Clippings, c. July 1926–c. August 1933. For more on Mary and Cherry Kearton, see Kearton, "Mary Was My Cleverest Pet."

73. Lady Clara, Chimpanzee, Clifton Zoo, 9 July 1923, Roy Vaughan Collection.

74. *Evening Post,* 18 May 1951, and *Bristol Evening Post,* 5 April 1958, in Newspaper Clippings, c. March 1936–c. July 1961; *New Observer,* 8 October 1971, in Newspaper Clippings, c. March 1963–c. October 1972.

75. Book Publishing Animal Magic, 1966–68 (File 1), BBC Written Archives Centre; "Episode 15," *Animal Magic,* 19 December 1962; "Episode 27," *Animal Magic,* 4 June 1963; "Episode 331," *Animal Magic,* 4 November 1975; "Episode 409: Animal Magic on the Inside," *Animal Magic,* 6 September 1983; "Johnny's Animal Magic," *Animal Magic,* 20 April 1984: all BBC Audio-Visual Archive.

76. *Bristol Evening World,* 11 December 1926, in Newspaper Clippings, c. July 1926–c. August 1933; Meeting 1 October 1856 and 7 November 1860, in General

Committee, April 1856–June 1861, 12, 189; Meeting 4 December 1961, in Council Minutes, February 1956–February 1966, 7; *Programme of the Midsummer Carnival,* 57; Vicars-Webb *et al., Guide to the Gardens,* 1928, 19.

77. *Bristol Zoo Book,* 1948, 50–51; *Bristol Zoo Book,* 1957, 48; Hartley, *Bristol Zoo Book,* c. 1973, 58; Vanderplank, *Official Guide to the Gardens,* 1937, 21; Vicars-Webb *et al., Guide to the Gardens,* 1928, 19.

78. Barrow, "Alligator's Allure," 127–29; Sax, "Basilisk and the Rattlesnake," 12.

79. *Bristol Evening World,* 18 February 1936, in Newspaper Clippings, c. September 1933–c. August 1937; *Bristol Evening Post,* 10 October 1963, in Newspaper Clippings, c. September 1961–c. April 1965.

80. Nichols, "Bristol Zoo Revisited."

81. *Bristol Evening Post,* 10 October 1963, in Newspaper Clippings, c. September 1961–c. April 1965.

82. Mike Colbourne, interview, Bristol, 16 March 2011; "Bachelor Is the Zoo's New 'Mum,'" *Nationwide,* c. 1977, BBC Audio-Visual Archive; *Nationwide,* c. 1978, BBC Audio-Visual Archive.

83. Ms. T. Hughes to Bristol Zoo, 30 May 1990, in Complimentary letters, c. 1990–c. 1995; *Daily Mail,* 27 July 1990, in the Colbourne Collection.

84. For a range of insightful essays around anthropomorphism, see Daston and Mitman, *Thinking with Animals.*

85. *Bristol Evening Post,* c. August 1984, in Newspaper Clippings, c. August 1976–c. November 1986.

86. *Daily Express,* 5 October 1978, in Newspaper Clippings, c. August 1976–c. November 1986; "Sanlam Gorilla Island," Bristol Zoo, accessed 22 January 2017, http://www.bristolzoo.org.uk/explore-the-zoo/gorilla-island.

87. "Baby Gorilla Opens Presents at Zoo on His First Birthday (with a Little Help from Mum)," *Daily Mail,* 27 September 2012, accessed 22 January 2017, http://www.dailymail.co.uk/news/article-2209253/Baby-gorilla-celebrates-birthday-zoo-slice-fruit-cake.html.

88. Benbow, "Death and Dying," 379–98; "Bristol Zoo Asiatic Lioness Moti Dies after Illness," BBC, 19 May 2010, accessed 22 January 2017, http://news.bbc.co.uk/local/bristol/hi/people_and_places/nature/newsid_8691000/8691970.stm; "Kemal the Lion at Bristol Zoo Has Died," 22 November 2012, accessed 22 January 2017, http://www.bristolpost.co.uk/kemal-lion-bristol-zoo-died/story-17393557-detail/story.html.

89. *Bristol Zoo Matters,* Winter 1994, 11; *Bristol Zoo Matters,* Summer 2001, 4.

90. Margaret W, interview, 31 October 2012.

91. *Western Daily Press,* 28 September 1974, in Newspaper Clippings, c. March 1973–c. April 1974; *Bristol Observer,* c. 1993, in Press cuttings, press releases, magazine articles about Bristol Zoo and zoos, and also polar bear publicity, c. 1990–c. 1993.

92. Richard Clarke, interview, 11 March 2013; Martin Davies Jones, interview, 20 March 2013; Brian Colenso, interview, 30 November 2012; Elephant Euthanasia

Protocol, n.d., Elephant Post-mortem Examination Update, 9 September 2012, and Elephant Euthanasia and Post-mortem, 10 September 2002, all in Elephant animal file, c. March–c. October 2002; Jo Gipps, interview, Bristol, 20 September 2012.

93. See, for instance, Kerr and Steinberg, *Mourning Diana.*

94. Condolence Books for Wendy the Elephant.

95. "Bristol Zoo Hasn't Been the Same Since Wendy Died," Facebook, accessed 6 October 2011, https://www.facebook.com/media/set/?set=0.2231945287&ref=mf.

96. *Bristol Zoo Matters,* Spring 1994, 1; *Bristol Zoo Matters,* Winter 2000, frontispiece; *Bristol Zoo Matters,* Winter 2002, frontispiece; *Bristol Zoo Matters,* Summer 2001, 2.

97. *Bristol Zoo Gardens,* 1987, 9–11; *Bristol Zoo Gardens Guidebook,* 2001, 12.

98. Braverman, *Zooland,* 62.

99. Correspondence from the children of G. T. Totham C. P. School, Essex, to Bristol Zoo, May 1990, and Correspondence from Mrs. S. Stirret to Mr. G. Greed, 10 July 1995, in Complimentary letters, complaints, and membership, c. 1990–c. 1995.

100. "Animal Phobia Courses at City Zoo," BBC, 6 November 2009, accessed 22 January 2017, http://news.bbc.co.uk/local/bristol/hi/people_and_places/nature/newsid_8346000/8346380.stm.

5. THE POWER OF NATURE, AND THE NATURES OF POWER

1. Meeting 26 September 1944, in General Committee and Executive Council, August 1944–January 1956, 2; "Night Vigil at Rosie's Bedside," [n.pub.] 19 August 1947, and *Times,* 19 August 1947 and 25 July 1960, in Newspaper Clippings, c. March 1936–c. July 1961.

2. Ingold, *Lines.*

3. Shaw, "Way with Animals," 7–9. For works that consider the agency of other animals—with varying degrees of success—see, for instance, Adams Martin, "When Sharks (Don't) Attack"; Flack, "In Sight, Insane"; Hribal, "Animals, Agency and Class"; L. Nash, "Agency of Nature"; Pearson, "Between Instinct and Intelligence"; Swart, *Riding High.*

4. Foltz, "Does Nature Have Historical Agency?"

5. Sack, *Human Territoriality;* Woods, "Fantastic Mr. Fox?," 199.

6. *London Dispatch and People's Political and Social Reformer,* 18 November 1838; *London Morning Post,* 9 March 1846.

7. *Bristol Evening World,* 26 March 1934 and 24 March 1934, in Newspaper Clippings, c. September 1933–c. August 1937.

8. Nick Hood, interview, Bristol, 25 February 2013; *Northern Times (Carnarvon, Western Australia),* 25 August 1949; Don Packham, interview, Bristol, 13 December 2010; Janet Rose, interview, Bristol, 23 November 2011. Also see *Times,* 2 July 1955, in Newspaper Clippings, c. March 1936–c. July 1961.

9. *Bristol Times and Felix Farley's Bristol Journal,* 17 June 1854; "Madame Juliet at Home," *Bristolian,* c. July 1914, Bristol Record Office.

10. *Bristol Evening World,* 20 June 1938; *Western Daily Press,* 20 June 1938; *Bristol Evening Post,* 28 February 1948; *Daily Mirror,* 20 June 1938; *Times,* 20 June 1938; *Bristol Evening Post,* 3 August 1938, and *Bristol Evening World,* 19 June 1938, all in Newspaper Clippings, c. March 1936–c. July 1961; Meetings 6 June 1932, 15 March 1933, 22 June 1938, and 7 October 1939, all in General Committee, January 1931–July 1944, 54, 80, 238, 274.

11. *Bristol Evening Post,* 12 November 1973, in Newspaper Clippings, c. March 1973–c. April 1974.

12. *Western Daily Press,* 4 June 1962, in Newspaper Clippings, c. September 1961–c. April 1965; Frank Wooff, interview, Torquay, 11 August 2011; Jennifer and Robert Bishop, interview, Bristol, 11 April 2011; *Bristol Evening Times,* 25 October 1931, in Newspaper Clippings, c. July 1926–c. August 1933; *Western Daily Press,* 26 July 1960, in Newspaper Clippings, c. March 1936–c. July 1961; Meeting 2 January 1929, in General Committee, January 1912–December 1930, 272; Meeting 4 November 1941, in General Committee, January 1931–July 1944, 29–31.

13. John Partridge and Alice Warren, email to the author, 28 January 2011.

14. *Daily Mirror,* 26 June 1928, in Newspaper Clippings, c. July 1926–c. August 1933; Brian Colenso, interview, Bristol, 30 November 2012; *Bristol Evening World,* 3 August 1935, in Newspaper Clippings, c. September 1933–c. August 1937; *Western Daily Press,* 18 March 1929, and *Times,* 16 November 1929, both in Newspaper Clippings, c. July 1926–c. August 1933; Meeting 17 July 1935, in General Committee, January 1931–July 1944, 148–49; *Times,* 20 June 1964, in Newspaper Clippings, c. September 1961–c. April 1965.

15. Hribal, *Fear of the Animal Planet,* 42; Ritvo, *Animal Estate,* 224–26.

16. Meeting 19 December 1840, in General Committee, April 1836–October 1848, 243; Meetings 4 February 1863 and 4 March 1863, in General Committee, June 1861–September 1866, 82, 87; *Daily Chronicle,* 11 February 1927, in Newspaper Clippings, c. July 1926–c. August 1933; Report on Elephant Wendy, c. 2002, and Wendy the Elephant, c. 2002, in Elephant animal file, c. March–c. October 2002.

17. Meeting 19 July 1933, in General Committee, January 1912–December 1930, 89–90; Meeting 20 December 1977, in Council Minutes, September 1967–December 1983, 125.

18. Meeting 20 September 1939, in General Committee, January 1931–July 1944, 271.

19. Meeting 3 November 1875, in General Committee, June 1861–September 1866, 294; *Bristol Evening Post,* 12 November 1973.

20. Hribal, *Fear of the Animal Planet,* 48.

21. *Western Daily Press,* 15 March 1929.

22. For more on the artificial insemination of female pandas, see Miller, *Nature of the Beasts,* 226.

23. *Bristol Times and Felix Farley's Bristol Journal,* 15 April 1854; "Rhesus Goes Back to the Temple," [n.pub.], c. 1934, and *Bristol Evening World,* 24 March 1934, in Newspaper Clippings, c. September 1933–c. August 1937.

24. *Bristol Evening World,* 11 August 1937, in Newspaper Clippings, c. September 1933–c. August 1937.

25. "Owl Ends His Self-Imposed Exile," BBC News, 11 October 2005, accessed 23 January 2017, http://news.bbc.co.uk/2/hi/uk_news/england/bristol/somerset/4330420.stm.

26. Plumwood, "Being Prey"; Walker, "Animals and the Intimacy."

27. *Bristol Times and Felix Farley's Bristol Journal,* 17 June 1854; *Traralgon Record* (Victoria) 24 March 1932.

28. *Bristol Evening World,* 20 June 1938; *Western Daily Press,* 20 June 1938; *Bristol Evening Post,* 28 February 1948; *Daily Mirror,* 20 June 1938; *Times,* 20 June 1938; *Bristol Evening Post,* 3 August 1938, and *Bristol Evening World,* 19 June 1938, all in Newspaper Clippings, c. March 1936–c. July 1961.

29. *Bristol Evening World,* 20 June 1938; *Daily Mirror,* 20 June 1938; *Bristol Evening Post,* 3 August 1938, all in Newspaper Clippings, c. March 1936–c. July 1961. Also see Hribal, *Fear of the Animal Planet,* 22–27.

30. Wood, "'Killing the Elephant.'"

31. Meeting 7 May 1930, in General Committee, January 1912–December 1930, 336; *Bristol Evening Post,* 14 April 1952, in Newspaper Clippings, c. March 1936–c. July 1961; G. R. Greed Report on attack by Orang-utan Henry on keeper Attwood, 30 October 1974, and Report to Commercial Union Assurance, 26 September 1974, in Attack by Orangutan Henry on Keeper Attwood, 8 May 1974–c. February 1975.

32. Carol Jacobs to Geoffrey Greed, 26 August 1981, H. H. Drew to Bristol Zoo, 29 August 1981, and Health and Safety Executive Report of an accident and/or dangerous occurrence and injuries sustained, 25 August 1981, in Correspondence Files, Reptile House incident, 25 August 1981; *Western Daily Press,* 27 August 1981; *Times,* 26 August 1981.

33. *Western Daily Press,* 27 January 1927, in Newspaper Clippings, c. July 1926–c. August 1933; *Western Daily Press,* 15 March 1929 and 18 March 1929.

34. *Bristol Times and Mirror,* 4 April 1931, in Newspaper Clippings, c. July 1926–c. August 1933. Also see Fuentes, "Monkey and Human Interconnections."

35. "A Grand Old Lady Who Likes Work," [n.pub.], c. 1950, in Newspaper Clippings, c. March 1936–c. July 1961.

36. *Woman and Home,* c. 1977, in Newspaper Clippings, c. August 1976–c. November 1986.

37. *Bristol Evening Post,* 11 June 1983, and *Bristol Evening Post,* 6 May 1983, in Newspaper Clippings, c. August 1976–c. November 1986.

38. Nagel, "What Is It Like?"

39. Weeratunge, *Elephant Gates,* 104; Wylie, *Elephant,* 43, 123.

40. *Western Gazette,* 18 June 1926, in Newspaper Clippings, c. July 1926–c. August 1933.

41. Ray Flay, interview, Bristol, 20 July 2012; Janet Rose, interview, Bristol, 23 November 2011; Meeting 16 July 1941, in General Committee, January 1931–July 1944), 332; Meeting 26 March 1958, in Council Minutes, February 1956–February 1966, 72.

42. Hart, "Elephant-Mahout Relationship." Also see Mitman, "Pachyderm Personalities."

43. Hediger, *Wild Animals in Captivity*, 47 (emphasis added). Also see Hosey, "Hediger Revisited."

44. *Times*, 27 March 1972.

45. Scott, "Racehorse as Protagonist," 63; Mike Colbourne, interview, Bristol, 16 March 2011.

46. See, for instance, *Western Daily Press*, 18 May 1971, in Newspaper Clippings, c. March 1963–c. October 1972.

47. Correspondence from A. Fishlock, 7 January 1987, in Polar Bear letters and press cuttings 1986–89; "An open letter to any member of Bristol Zoo Action, if the cap fits," in Keeling, *Bristol Book;* Correspondence from Ms. J. Powell to Mr. Greed, 22 September 1993, in Complimentary letters, complaints and membership from members of the public 1990–95; Correspondence from Mrs. Maureen Bull, 6 February 1992, in Polar Bear letters and press cuttings 1986–89; Correspondence from Mrs. A Tanner, 11 February 1987, in Polar Bear letters and press cuttings 1986–89.

48. Meeting 3 January 1979, Animal Sub-committee, January 1978–February 1988.

49. Horseman, *Captive Polar Bears.*

50. Baratay and Hardouin-Fugier, *Zoo,* 274; Braitman, *Animal Madness.*

51. *Newcastle Weekly Courant,* 16 February 1889.

52. *Bristol Mercury and Daily Post,* 26 October 1894; Bristol Zoo Gardens Historical Footage, c. 1930–c. 1995. On the use of historical film footage, see Flack, "Capturing the Beasts."

53. Zoo Check polar bear research, c. November 1989–c. January 1990.

54. *Bristol Evening Post,* 22 September 1986; *Mail on Sunday,* 28 September 1986, in Newspaper Clippings, August 1976–c. November 1986.

55. *Bristol Journal,* 9 October 1986, in Newspaper Clippings, August 1976–c. November 1986; "Polar Bears," *BBC News,* 29 September 1986, BBC Audio-Visual Archive.

56. E. M. Wilson, 13 March 1989, in Polar Bear letters and press cuttings, 1986–89; *Mail on Sunday,* 12 March 1989; *Sunday Times,* 12 March 1989; *Daily Star,* 13 March 1989; *Mail on Sunday,* 19 March 1989; *Daily Express,* c. 1989; *Western Daily Press,* 13 March 1989 in Polar Bear letters and press cuttings, 1986–89.

57. Correspondence from the Public, in Polar Bear letters and press cuttings, 1986–89.

58. Sue Clayton, Southampton, 1 October 1986, P. V. Barber, Cowes, Isle of Wight, n.d., and Correspondence from Children, all in Polar Bear letters and press cuttings, 1986–89.

59. Paying visitors, 1943–2011, in *Annual Report of the BCWEZS, 1940–1989.*

60. "Polar Bears Update," *Zoo Check News,* 7, July 1989, 16.

61. *Mail on Sunday,* 26 March 1989; Press Release from the National Federation of Zoological gardens of Great Britain and Ireland, 10 November 1989, and *Eve-*

ning Chronicle, 6 March 1990, both in Press cuttings, press releases, magazine articles about Bristol Zoo and zoos, and also polar bear publicity, 1990–93; Ames, "Management and Behaviour"; *Daily Mail,* 18 December 1989; *Today,* 6 December 1989; Meeting 26 April 1990, Council Meetings, January 1984–December 1994; Redshaw, *Study of the Effects; Western Daily Press,* 6 December 1989.

62. *Mail on Sunday,* 5 August 1990, and *Sunday Express,* 5 August 1990, both in Press cuttings, press releases, magazine articles about Bristol Zoo and zoos, and also polar bear publicity, 1990–93.

63. *Life and Times,* 15 September 1992, *Prima Magazine,* December 1992, *Daily Telegraph,* 8 April 1991, *Guardian,* 17 July 1992, *Daily Telegraph,* 16 June 1989, and *Independent on Sunday,* 20 October 1991, all in Press cuttings, press releases, magazine articles about Bristol Zoo and zoos, and also polar bear publicity, 1990–93.

64. *Western Daily Press,* 10 July 1991; *Daily Star,* 7 March 1991; *Bristol Evening Post,* 28 March 1990.

65. *Annual Report of the BCWEZS,* 1992, 7; BCWEZS Press Release, 20 January 1992, in Polar Bear letters and press cuttings, 1986–89; Geoffrey Greed, interviews, Bristol, 13 December 2010, and Clevedon, 29 July 2013.

66. Meeting 26 July 1993, in Council Meetings, January 1984–December 1994, 142.

67. Bristol Zoo Action, Factsheet No. 1: Welfare, c. 1993, Bristol Zoo Action Press Release, 18 September 1993, *Bristol Observer,* 16 July 1993, all in Polar bears letters, zoo check newsletter, letters from other zoos, c. 1987–c. 1899.

68. "The Don't Care Bears," *Punch,* 28 April 1989, 36–39.

69. *Creature Comforts,* 1991, Aardman Animations, Bristol.

70. Foucault, *Power/Knowledge,* 98.

71. On the polar bear as a heritage species see Coates, "Creatures Enshrined."

72. *Mail on Sunday,* 5 October 1986, in Newspaper Clippings, c. August 1976–c. November 1986; R. J. Wheather, Royal Zoological Society of Scotland, to Brigadier S. E. M. Goodall, 5 October 1987, in Polar bears letters, zoo check newsletter, letters from other zoos, c. 1987–c. 1999.

73. Meetings 15 May 1935, 12 June 1935, and 16 October 1935, all in General Committee, January 1912–December 1930, 144, 146, 156.

74. Transcript of Radio Bristol Interview with Julian Hector, 19 March 1989, in Correspondence with Slimbridge Wildfowl and Wetlands Trust, November 1982–July 1992; Public Relations Correspondence, 1989.

75. Meeting 26 April 1990, in Council Meetings, January 1984–December 1994, 93; "Zoo Check Report," *BBC Points West,* 23 March 1993, BBC Audio-Visual Archive.

76. Meeting 4 April 1866 and 4 July 1866, in General Committee, June 1861–September 1866, 221; Meeting 4 November 1931, in General Committee, January 1912–December 1930, 29–31; BCWEZS whistle-blowing policy, 2007.

6. THE FACES OF THE BEASTS

1. "Home," World Association of Zoos and Aquariums (WAZA), accessed 22 January 2017, http://www.waza.org/en/site/zoos-aquariums. This figure is likely to be a gross underestimation of the total number of visitors frequenting all zoos and aquariums of the world. The figure of c. 700 million relates only to the 1,300 sites that are in some way associated with WAZA.

2. Baratay and Hardouin-Fugier, *Zoo,* 13.

3. Fudge, *Animal,* 28–32, 133, 160.

4. Berger, "Why Look at Animals?," 24.

5. Grinfield, *Our Clifton Zoo,* 28–29; Correspondence from Angela Valtingham, c. 1993, Memories of Alfred: John Garland, 28 July 1993, Memories of Alfred: Tony Osborne, 9 November 1996, Memories of Alfred: Cynthia Pearson, 20 July 1993, all in Alfred file, BMAG.

6. Leubke et al., *Global Climate Change,* 5. It must be said, however, there is far from a consensus about how effective zoos actually are in promoting such awareness and compassion toward the rest of the natural world.

7. Meeting 4 August 1971, in Meetings of the Animal Sub-committee, 27 September 1967–28 December 1977, 70; *Bristol Evening Post,* 23 July 1971, in Newspaper Clippings, c. March 1963–c. October 1972.

8. "Facts and Figures," RSPCA, accessed 22 January 2017, https://media.rspca.org.uk/media/facts; "Pet Population 2016," Pet Food Manufacturers' Association (PFMA), accessed 22 January 2017, http://www.pfma.org.uk/pet-population-2016; "Zoos and Aquaria—UK," Born Free, accessed 22 January 2017, http://www.bornfree.org.uk/campaigns/zoo-check/zoos/zoos-in-the-uk/.

9. "Farm Animal Statistics: Slaughter Totals," Humane Society of the United States, accessed 22 January 2017, http://www.humanesociety.org/news/resources/research/stats_slaughter_totals.html?; "Pet Industry Market Size and Ownership Statistics," American Pet Products Association (APPA), accessed 22 January 2017, http://www.americanpetproducts.org/press_industrytrends.asp; "Critics Question Zoos' Commitment to Conservation," National Geographic News, 13 November 2003, accessed 22 January 2017, http://news.nationalgeographic.com/news/2003/11/1113_031113_zoorole.html.

10. "Living Planet Report 2016," WWF, accessed 22 January 2017, http://assets.wwf.org.uk/custom/lpr2016/.

11. Meeting 18 April 1966, in Council Minutes, March 1966–August 1967, 6.

12. National Wildlife Conservation Park Masterplan, January 2001.

Bibliography

ARCHIVAL SOURCES

Archive of the Bristol Zoological Society (BZS)

Minute Books

Animal Committee, c. 1985–c. 1986 (TEMP/402).
Animal Sub-committee, September 1967–December 1977 (TEMP/2716).
Animal Sub-committee, January 1978–February 1988 (TEMP/2717).
Council Meetings, January 1984–December 1994.
Council Minutes, February 1956–February 1966.
Council Minutes, March 1966–August 1967.
Council Minutes, September 1967–December 1983.
Education, Research, Conservation and Scientific Committee, c. June 1986–
    c. December 1995 (TEMP/415).
General Committee and the Executive Council, August 1944–January 1956.
General Committee and the Finance Committee, October 1866–September 1880
    (TEMP/2693).
General Committee and the Finance Committee, November 1880–March 1893
    (TEMP/2694).
General Committee, April 1836–October 1848 (TEMP/2689).
General Committee, April 1856–June 1861 (TEMP/2691).
General Committee, June 1861–September 1866 (TEMP/2692).
General Committee, April 1893–December 1911 (TEMP/2695).
General Committee, January 1912–December 1930 (TEMP/2696).
General Committee, January 1931–July 1944.
General Meetings, July 1835–April 1854 (TEMP/2688).
Special and later General Committee, May 1836–April 1856 (TEMP/2690).
Zoology Sub-committee and Fetes Committee, May 1836–September 1850
    (TEMP/2714).

Correspondence Files

Acquisition of an elephant from Balasore, India, c. 1912–c. 1913 (TEMP/1362).

Attack by orangutan Henry on Keeper Attwood, May 1974–c. February 1975 (TEMP/394).

Complimentary letters from visitors praising Bristol Zoo, c. 1974–c. 1978 (TEMP/1233).

Complimentary letters, complaints and membership from members of the public, c. 1990–c. 1995 (TEMP/3155).

Cross, Miss Hilary, September 2011.

Davies, Maxine, January 2012.

Enquiries regarding gifts/donations of animals, or requests for hair/feathers, c. 1912 (TEMP/1361).

Miscellaneous correspondence, c. 1911–c. 1912 (TEMP/1357).

Misra, Betty, August 2011.

Offers of animals from the general public, c. 1910–c. 1912 (TEMP/1345).

Other UK zoos, c. 1911 (TEMP/1348).

Perks, Philippa, July 2011.

Reptile House incident, August 1981.

Share certificates, letters to animal dealers, and opticians, c. 1893–c. 1894 (TEMP/1329).

Slimbridge Wildfowl and Wetlands Trust, November 1982-July 1992 (TEMP/890).

Subscriptions, complimentary entrance tickets and general enquiries, c. 1911 (TEMP/1341).

Vets and professors of livestock, c. September 1970–c. December 1982 (TEMP/459).

Williams, Mrs. Mavis, July 2011.

Zoo Check polar bear research, c. November 1989–c. January 1990.

Newspaper Clipping Albums and Folders

Newspaper Clippings, c. July 1926–c. August 1933 (TEMP/2629).

Newspaper Clippings, c. September 1933–c. August 1937 (TEMP/2630).

Newspaper Clippings, c. March 1936–c. July 1961 (TEMP/2631).

Newspaper Clippings, c. September 1961–c. April 1965 (TEMP/2632).

Newspaper Clippings, c. February 1963–c. December 1964 (TEMP/2633).

Newspaper Clippings, c. March 1963–c. October 1972 (TEMP/2634).

Newspaper Clippings, c. March 1973–c. April 1974 (TEMP/2635).

Newspaper Clippings, c. August 1976–c. November 1986 (TEMP/2636).

Film (No TEMP Numbers)

Bristol Zoo Gardens Historical Footage, c. 1930–c. 1995.

Animal Department Specialist Reports
Shiva, Lioness 11399, ZIMS Report (Generated 5 March 2011).
Specimen Report for BRISTOL/11893, ARKS Report (Generated 3 August 2011).

Oral History Interviews
Bishop, Jennifer and Robert, Bristol, 11 April 2011.
Bland, Gordon, Bristol, 16 November 2011.
Clarke, Richard, Bristol, 11 March 2013.
Colbourne, Mike, Bristol, 16 March 2011.
Colenso, Brian, Bristol, 30 November 2012.
Davies Jones, Martin, Bristol, 20 March 2013.
Flay, Ray, Bristol, 20 July 2012.
Floyd, Peter, Bristol, 21 November 2012.
Gipps, Jo, Bristol, 20 September 2012.
Greed, Geoffrey, Bristol, 13 December 2010.
Greed, Geoffrey, Clevedon, 29 July 2013.
Hewer, Richard, Bristol, 5 July 2013.
Hood, Nick, Bristol, 25 February 2013.
Packham, Don, Bristol, 13 December 2010.
Packham, Don, Bristol, 29 January 2013.
Partridge, John, Bristol, 13 June 2011.
Rose, Janet, Bristol, 23 November 2011.
W., Margaret, Bristol, 31 October 2012.
Wooff, Frank, Torquay, 11 August 2011.

Personal Collections (No TEMP Numbers)
The Colbourne Collection.
The Roy Vaughan Collection.

Miscellaneous Records
Adoption Scheme accounts, 1940–41 (TEMP/2819).
Adoption Scheme receipts, 1940–41 (TEMP/1323/1).
Animal Reports, 1923–69 (TEMP/2823–2854).
Animal Reports, 1970–93 (TEMP/144–157).
Ansell, C. M., "*Pasteurella Pseudotuberculosis* in Bristol Zoo," c. October 1969–
    c. October 1970 (TEMP/33).
Annual Stock List, 2010.
BCWEZS whistle-blowing policy, November 2007.
Cat complex 1977, c. 1977 (TEMP/3148).
Condolence books for Wendy the elephant, 2002 (TEMP/3000).
Elephant animal file, c. March–c. October 2002.

Mata Atlantica (Zona Brasil) Dutton Merrifield campaign brief, February 2002 (TEMP/2971).

National Wildlife Conservation Park Masterplan, January 2001.

Nichols, Peter, "Bristol Zoo Revisited," c. 1993.

Polar Bear letters and press cuttings, c. 1986–c. 1989 (TEMP/2996).

Polar bears letters, zoo check newsletter, letters from other zoos, c. 1987–c. 1999 (TEMP/2997).

Post Mortem Reports, 1928 (TEMP/1543).

Post Mortem Reports, 1976–1988 (TEMP/1578).

Press cuttings, press releases, magazine articles about Bristol Zoo and zoos, and also polar bear publicity, c. 1990–c. 1993 (TEMP/2999).

Stock Book, 1935–37 (TEMP/2820).

Twilight World, c. 1996 (TEMP/2983).

Twilight World, c. 1996–c. 1999 (TEMP/2982).

BBC Written Archives Centre Files

Book Publishing Animal Magic, 1966–68 (File 1) (R43/745/1).

Bristol Museum & Art Gallery (BMAG)

Alfred file, c. 1930–c. 2010.

Report of the Museum and Art Gallery Committee for the Year Ending 30 September 1931. Bristol: BMAG, 1931.

Bristol Record Office (BRO)

"An Act for extending the Powers of Sale and Exchange contained in the Marriage Settlement of Francis Adams the younger, Esquire; and for other Purposes," 17 July 1837, Anno Primo Victoria Reginae, Cap. 43 (private bill) (BRO/43041/2).

"Madam Juliet at Home," *Bristolian,* c. July 1914, Bristol Record Office, BRO/40156(16).

Melville, Frederick George, "Visit of St Mary Redcliff National Schools to the Zoo Gardens," 27 June 1879 (BRO/40477/HM/3).

Bristol Reference Library

The Bristol Presentment: or Bill of Entry (BPBE), 35–62, 1853–80.

Bristol Zoo Newspaper Cuttings, 1932–85 (909).

Bristol Zoo Newspaper Cuttings, 1986–2011 (909).

Port of Bristol HM Customs Bill of Entry (PBHMCBE), 1–13, 1881–93.

The National Archives, Kew

Export of penguins from South Georgia for the Bristol Zoo, 12 January 1937–c. May 1938 (CO78:206/11).

PUBLISHED MATERIALS

Official Publications of the Bristol, Clifton and West of England Zoological Society (BCWEZS), now known as the Bristol Zoological Society.

Annual Reports. Bristol: BCWEZS, 1836–2011.

Annual Review and Financial Statements. Bristol: BCWEZS, 2000 and 2006.

Bristol Zoo Book and Official Guide. Bristol: BCWEZS, 1948.

Bristol Zoo Book and Official Guide. Bristol: BCWEZS, 1952.

Bristol Zoo Book and Official Guide. Bristol: BCWEZS, 1954.

Bristol Zoo Book and Official Guide. 2nd ed. Bristol: BCWEZS, 1954.

Bristol Zoo Book and Official Guide. Bristol: BCWEZS, 1957.

Bristol Zoo Book and Official Guide. Bristol: BCWEZS, 1964.

Bristol Zoo Book and Official Guide. Bristol: BCWEZS, 1967.

Bristol Zoo Gardens: The Exciting Zoo Set in Twelve Acres of Beautiful Gardens. Bristol: BCWEZS, 1987.

Bristol Zoo Gardens Guidebook with Map. Bristol: BCWEZS, 2001.

Bristol Zoo Gardens: A New Zoo Every Day. Bristol: BCWEZS, 1992.

Bristol Zoo Gardens Official Guide. Bristol: BCWEZS, 2006.

Bristol Zoo Gardens Official Guide. Bristol: BCWEZS, 2007.

Bristol Zoo Gardens Official Guide. Bristol: BCWEZS, c. 2009.

Bristol Zoo Matters. Bristol: BCWEZS, c. 1993–c. 2010.

Catalogue of the Animals, Birds &c in the Bristol and Clifton Zoological Gardens, with short descriptive notices of their organisation and habits, as well as in a state of nature as in captivity. Bristol: Henry and Alfred Hill, 1843.

Hartley, J. C. *Bristol Zoo Book and Official Guide.* Bristol: BCWEZS, c. 1965.

———. *Bristol Zoo Book and Official Guide.* Bristol: BCWEZS, c. 1973.

Illustrated Official Guide to the Clifton Zoological Gardens. Bristol: J. W. Arrowsmith, 1894.

Illustrated Official Guide to the Gardens of the BCWEZS. Bristol: Imperial Publishers, c. 1924–25.

Knocker, Fredc W. *Illustrated Official Guide to the Gardens of the BCWEZS.* Bristol: Fred G. Melville, 1899.

———. *Illustrated Official Guide to the Gardens of the BCWEZS.* Bristol: George du Boistel & Co, 1903.

———. *Illustrated Official Guide to the Gardens of the BCWEZS.* Bristol: George du Boistel & Co, 1904.

———. *Illustrated Official Guide to the Clifton Zoological Gardens.* Bristol: George du Boistel & Co, 1909.

———. *Illustrated Official Guide to the Clifton Zoological Gardens.* Bristol: B. H. Matthews, 1910.

London Zoo Guide. London: Zoological Society of London, 1958.

Phillips, J. Le H. *The Gardens and Menagerie of the BCWEZS*. Bristol: C. T. Jefferies and Sons, c. 1882.

Programme for Bristol Hospitals Summer Carnival. Bristol: BCWEZS, 1932.

Programme of the Midsummer Carnival. Bristol: BCWEZS, 1928.

Programme for the Zoo Carnival. Bristol: BCWEZS, 1919.

Vanderplank, F. L. *Official Guide to the Gardens and Aquarium of the BCWEZS*. Bristol: BCWEZS, 1937.

Vanderplank, F. L., and Gwen Woolls, *Bristol Zoo Book and Official Guide*. Bristol: BCWEZS, 1938.

Vicars-Webb, H. *et al. The Bristol, Clifton and West of England Zoological Society Guide to the Gardens, Beasts, Birds and Fishes*. Bristol: BCWEZS, 1926.

———. *The Bristol, Clifton and West of England Zoological Society Guide to the Gardens, Beasts, Birds and Fishes*. Bristol: BCWEZS, 1928.

———. *The Bristol, Clifton and West of England Zoological Society Guide to the Gardens, Beasts, Birds and Fishes*. Bristol: BCWEZS, 1931.

Other Monographs and Journal Articles

"Address." *West of England Journal of Science and Literature* 1, no. 1 (1835): i–vii.

Ames, Alison. "Management and Behaviour of Polar Bears in Captivity." BA diss., Universities Federation for Animal Welfare, October 1990.

Aristotle. *De Anima*. Translated by Christopher Shields. Oxford: Clarendon Press, 2016.

———. *Aristotle's History of Animals, in Ten Books*. Translated by Richard Cresswell. London: Henry Bohn, 1862.

Bartlett, Abraham Dee. *Wild Animals in Captivity; Being an Account of the Habits, Food, Management and Treatment of the Beasts and Birds at the "Zoo" with Reminiscences and Anecdotes*. London: Chapman and Hall, 1899.

Bristol: For U.S. Armed Forces in U.K. London: Spottiswoode, Ballantyne & Co, n.d.

Bristol Institution 13th Annual Meeting, 1836, with a list of all the papers read before the Philosophical and Literary Society, and of the honorary members and associates. To the whole is prefixed a general memoir of the Institution. 1836.

Bristol Museum Annual Reports 1891–1907. Bristol: BMAG.

Bruyninckx, Hans. "Majority of Europe's Species Have Unfavourable Conservation Status." *Parliament Magazine,* 4 June 2015.

Buckland, Francis Trevelyan. *Curiosities of Natural History*. London: 1857.

Cicero, Marcus Tullius. *Cicero: In Twenty-Eight Volumes. XIX De natura deorum: Academica*. Translated by H. Rackham. Cambridge, MA: Harvard University Press, 1967.

Darwin, Charles R. *The Descent of Man, and Selection in Relation to Sex*. London: John Murray, 1871.

———. *The Expression of the Emotions in Man and Animals: With Photographic and Other Illustrations.* London: John Murray, 1872.

———. *On the Origin of Species, by Means of Natural Selection: or, The Preservation of Favoured Races in the Struggle for Life.* London: John Murray, 1859.

"The Don't Care Bears." *Punch,* 28 April 1989.

du Chaillu, Paul B. *Explorations and Adventures in Equatorial Africa.* London: John Murray, 1861.

Fiennes, R. N. T. W. "Zoo Medicine and Human Medicine." *New Scientist,* no. 336 (16 March 1961): 674–76.

Fossey, Dian. *Gorillas in the Mist.* Boston, MA: Houghton Mifflin, 1983.

Grinfield, Florence G. *Our Clifton Zoo, and the Folks We Meet There: A Semi-Comic Description in Prose and Verse.* Weston-Super-Mare: J. R. Walters, "Gazette" Office, 1894.

Hagenbeck, Carl. *Beasts and Men: Being Carl Hagenbeck's Experiences for Half a Century among Wild Animals.* An abridged translation by Hugh S. R. Elliot and A. G. Thacker; with an introduction by P. Chalmers Mitchell. London: Longmans, Green, 1909.

Halloran, Patricia O'Connor. "A Bibliography of References to Diseases of Wild Mammals and Birds." *American Journal of Veterinary Research* 16, Supplement (1955): 1–465.

Hartman, R. *Les singes anthropoïdes et leur organisation comparée à celle de l'homme.* Paris: F. Alcan, 1884.

Hediger, H. *Man and Animal in the Zoo: Zoo Biology.* Translated by Gwynne Vevers and Winwood Reade. London: Routledge and Kegan Paul, 1969.

———. *Wild Animals in Captivity: An Outline of the Biology of Zoological Gardens.* Translated by G. Sircom. London: Butterworth Scientific, 1950.

Hill, Leonard. *Sunshine and the Open Air: Their Influence on Health with Special Reference to the Alpine Climate.* London: Edward Arnold, 1925.

Hornaday, W. T. "Gorillas, Past and Present." *Bulletin: New York Zoo Society* 18, no. 1 (1915): 1181–85.

Horsman, Paul V. *Captive Polar Bears in UK and Ireland Zoochotic Report.* N.p.: Zoo Check, 1988.

Huxley, Thomas. *Evidence as to Man's Place in Nature.* London: Williams and Norgate, 1864.

International Union for Conservation of Nature and Natural Resources Policy Statement on Captive Breeding. IUCN, 1 September 1986.

Jones, W. M. *Jarge Balsh at Bristol Zoo.* London: Herbert Jenkins, 1934.

Kearton, Cherry. "Mary Was My Cleverest Pet." *Zoo* 9, no. 1 (February 1937): 18–19.

Knight, Maxwell. *Animals after Dark.* London: Routledge and Kegan Paul, 1956.

Leubke, Jerry F., and others. *Global Climate Change as Seen by Zoo and Aquarium*

Visitors: The Climate Literacy Zoo Education Report Final Report. Chicago: Chicago Zoological Society, May 2012.

Linné, Carl von. *Systema naturæ, sive regna tria naturæ systematice proposita per classes, ordines, genera, & species.* 1735.

Lloyd Morgan, C. "On the Study of Animal Intelligence." *Mind: A Quarterly Review of Psychology and Philosophy* 11, no. 42 (1886): 174–85.

Loudon, J. C. *The Suburban Gardener, and Villa Companion: Comprising the Choice of a Suburban or Villa Residence, or of a Situation on Which to Form One; the Arrangement and Furnishing of the House; and the Laying Out, Planting, and General Management of the Garden and Grounds; the Whole Adapted for Grounds from One Perch to Fifty Acres and Upwards in Extent; and Intended for the Instruction of Those Who Know Little of Gardening and Rural Affairs, and More Particularly for the Use of Ladies. By J.C. Loudon, . . . Illustrated by Numerous Engravings.* London: J. C. Loudon, 1838.

"Only Six Left Alive." *Zoo* 1, no. 3 (August 1936): 52–54.

Orr, J. B. *Food Health and Income.* London: Macmillan, 1936.

Orwell, George. *Animal Farm.* London: Secker & Warburg, 1949.

Owen, Richard. *Memoir on the Gorilla (Troglodytes Gorilla, Savage).* London: Taylor and Francis, 1865.

———. "On the Extent and Aims of a National Museum of Natural History (Excerpt) (London, 1862)." In *The Emergence of the Modern Museum: An Anthology of Nineteenth-Century Sources,* edited by Jonah Siegel, 234–38. Oxford: Oxford University Press, 2008.

"Polar Bears Update." *Zoo Check News* 7. Horsham: Zoo Check, July 1989.

Redshaw, Margaret E. *A Study of the Effects of Environmental Enrichment on Captive Polar Bears (Thalarctos maritimus).* Bristol: BCWEZS, July 1989.

Tomlinson, Charles. *A Rudimentary Treatise on Warming and Ventilation: Being a Concise Exposition of the General Principles of the Art of Warming and Ventilating Domestic and Public Buildings, Mines, Lighthouses, Ships . . .* London: Virtue Brothers, 1864.

Tyson, Edward. *Orang-outang, sive Homo sylvestris: or, The anatomy of a pygmie compared with that of a monkey, an ape and a man. To which is added, A philological essay concerning the pygmies, the cynocephali, the satyrs, and sphinges of the ancients.* London, 1699.

Webb, W. T. *Webb's New Guide to Bristol, Clifton, the Zoological Gardens . . . : Also Notices of the Remarkable Places round Bristol, Interspersed with Anecdotes and Historical Information.* London, 1866.

Xenophon. *Memorabilia* of Socrates. Translated by R. D. C. Robbins. Andover: William H. Wardwell, 1848.

Newspapers and Periodicals

Atalanta (London)

Australian Women's Weekly (Sydney)

Barrier Miner (Broken Hill, New South Wales)
Birmingham Daily Post
Bristol Evening Post
Bristol Evening World
Bristol Mercury
Bristol Mercury and Daily Post
Bristol Observer
Bristol Times and Felix Farley's Bristol Journal
Bristol Times and Mirror
Clifton Chronicle & Directory: A Weekly Register of Local Intelligence, Arrivals, Removals, Departures, Amusements, Etc.
Daily Chronicle (London)
Daily Express
Daily Mail and Mail on Sunday (London)
Daily Mirror (London)
Daily Sketch (Manchester)
Ellesmere Guardian
Evening Chronicle (London)
Felix Farley's Bristol Journal
Free Lance-Star (Fredericksburg, VA)
Glasgow Herald
Hartford (CT) Courant
Independent on Sunday
Jackson's Oxford Journal
Life and Times
London Dispatch and People's Political and Social Reformer
Ludington (MI) Daily News
Manchester Guardian
Manchester Times and Gazette
Melbourne Argus
Milwaukee Journal
Milwaukee Sentinel
Morning Post (London)
Newcastle Weekly Courant
New Observer
New Yorker
Northern Liberator and Champion (Newcastle)
Northern Times (Carnarvon, Western Australia)
Pittsburgh Press
Prima Magazine
Royal Cornwall Gazette Falmouth Packet, Cornish Weekly News, & General Advertiser
Sheffield & Rotherham Independent

Spokane Daily Chronicle
Standard (London)
Sun (London)
Sunday Express (London)
Sydney Morning Herald
Telegraph (London)
Times (London)
Today (London)annual review
Traralgon Record (Victoria)
Western Daily Press (Bristol)
York Herald

British Broadcasting Corporation Audio-Visual Archive (BBC)
"Episode 15." *Animal Magic.* 19 December 1962 (NB59495B).
"Episode 27." *Animal Magic.* 4 June 1963 (NB59498J).
"Episode 331." *Animal Magic.* 4 November 1975 (NB59495B).
"Episode 409: Animal Magic on the Inside." *Animal Magic.* 6 September 1983
  (NB53311W).
"Johnny's Animal Magic." *Animal Magic.* 20 April 1984 (NB5C433S).
Nationwide. c. 1977 (C:LCA3750D).
Nationwide. c. 1978 (B:LCA3805T).
"New Aquarium." *BBC Points West.* c. 1987 (C:RWRB927K).
"Polar Bears." *BBC News.* 29 September 1986 (UWR178).
"Polar Bears." *BBC Points West.* 20 January 1992 (BWR147).
World Zoos: Bristol. 23 October 1962.
"Zoo Check Report." *BBC Points West.* 23 March 1993 (BNR272).

British Pathé Newsreels
"Barless Cage," c. 1950.
"Zoo Cubs aka New Arrivals at Bristol Zoo." c. 1968.

Miscellaneous Broadcasts and Film
Creature Comforts. c. 1989, Aardman Animations, Bristol.

Secondary Works
Acampora, Ralph R., ed. *Metamorphoses of the Zoo: Animal Encounter after Noah.*
  Lanham, MD: Lexington Books, 2010.
———. "Zoos and Eyes: Contesting Captivity and Seeking Successor Practices."
  Society & Animals 13, no. 1 (2005): 59–88.
Adams, Carol, and Josephine Donovan, eds. *Animals and Women: Feminist Theo-
  retical Explorations.* Durham, NC: Duke University Press, 1995.
Adams Martin, Jennifer. "When Sharks (Don't) Attack: Wild Animal Agency in
  Historical Narratives." *Environmental History* 16, no. 3 (2011): 451–55.

Adeney Thomas, Julia. "AHR Forum: Comment: Not Yet Far Enough." *American Historical Review* 117, no. 3 (2012): 794–803.

Alberti, Samuel J. M. M., ed. *The Afterlives of Animals: A Museum Menagerie.* Charlottesville: University of Virginia Press, 2011.

Alford, B. W. E. "Wills Family (*per.* c.1786–1928)." *Oxford Dictionary of National Biography.* Oxford University Press, 2004. http://www.oxforddnb.com/view/article/58453.

Allen, Garland E. *Life Science in the Twentieth Century.* London: John Wiley & Sons, 1975.

Altick, Richard D. *The Shows of London.* Cambridge, MA: Belknap Press of Harvard University Press, 1978.

Amato, Sarah. "The White Elephant in London: An Episode of Trickery, Racism and Advertising." *Journal of Social History* 43, no. 1 (2009): 31–66.

Anderson, Kay. "Culture and Nature at the Adelaide Zoo: At the Frontiers of 'Human' Geography." *Transactions of the Institute of British Geographers,* n.s., 20, no. 3 (1995): 275–94.

Anderson, Marla V., and Antonia J. Z. Henderson. "Pernicious Portrayals: The Impact of Children's Attachment to Animals of Fiction on Animals of Fact." *Society & Animals* 13, no. 4 (2005): 297–314.

Anderson, Virginia DeJohn. *Creatures of Empire: How Domestic Animals Transformed Early America.* Oxford: Oxford University Press, 2004.

Armstrong, Phillip. "The Post-Colonial Animal." *Society & Animals* 10, no. 4 (2002): 413–20.

Asma, Stephen T. *Stuffed Animals and Pickled Heads: The Culture and Evolution of Natural History Museums.* Oxford: Oxford University Press, 2001.

Auerbach, Jeffrey A. *The Great Exhibition of 1851: A Nation on Display.* New Haven: Yale University Press, 1999.

Axelsson, Tony, and Sarah May. "Constructed Landscapes in Zoos and Heritage." *International Journal of Heritage Studies* 14, no. 1 (2008): 43–59.

Bailey, Peter. *Leisure and Class in Victorian England: Rational Recreation and the Contest for Control, 1830–1885.* London: Routledge and Kegan Paul, 1978.

———. "The Politics and Poetics of Modern British Leisure: A Late Twentieth-Century Review." *Rethinking History: The Journal of Theory and Practice* 3, no. 2 (1999): 131–75.

Ballantyne, Tony. *Orientalism and Race: Aryanism and the British Empire.* Houndmills, Basingstoke, Hampshire: Palgrave Macmillan, 2002.

Bantock, Anton. *The Last Smyths of Ashton Court (Part II), from Their Papers, 1880–1900.* Bristol: Malago Society, 1998.

Barad, Karen. "Agential Realism: Feminist Interventions in Understanding Scientific Practices." In *The Science Studies Reader,* edited by Mario Biagioli, 1–11. London: Routledge, 1999.

Baratay, Eric, and Elisabeth Hardouin-Fugier. *Zoo: A History of Zoological Gardens in the West.* Translated by Oliver Welsh. London: Reaktion, 2002.

Barker, Kathleen. *Bristol at Play: Five Centuries of Live Entertainment.* Bradford-on-Avon: Moonraker Press, 1976.

Barker, W. R. *The Bristol Museum and Art Gallery: The Development of the Institution during a Hundred and Thirty-Four Years: 1772–1906.* Bristol: J. W. Arrowsmith, 1906.

Barrington-Johnson, J. *The Zoo: The Story of London Zoo.* London: Robert Hale, 2005.

Barrow, Mark V., Jr. "The Alligator's Allure: Changing Perceptions of a Charismatic Creature." In *Beastly Natures: Animals, Humans and the Study of History,* edited by Dorothee Brantz, 127–52. Charlottesville: University of Virginia Press, 2010.

———. *Nature's Ghosts: Confronting Extinction from the Age of Jefferson to the Age of Ecology.* Chicago: University of Chicago Press, 2009.

Beardsworth, Alan, and Alan Bryman. "The Wild Animal in Late Modernity: The Case of the Disneyisation of Zoos." *Tourist Studies* 1, no. 1 (2001): 83–104.

Bekoff, Marc. *The Emotional Lives of Animals: A Leading Scientist Explores Animal Joy, Sorrow and Empathy—and Why They Matter.* Novato, CA: New World Library, 2007.

———. "Increasing Our Compassion Footprint: The Animals' Manifesto." *Zygon* 43, no. 4 (2008): 771–81.

———. *Minding Animals: Awareness, Emotions, and Heart.* Oxford: Oxford University Press, 2002.

Beinart, William. "African History and Environmental History." *African Affairs* 99, no. 395, Centenary Issue: A Hundred Years of Africa (April 2000): 269–302.

———. "Empire, Hunting and Ecological Change in Southern and Central Africa." *Past and Present* 128 (1990): 162–86.

Benbow, S. Mary P. "Death and Dying at the Zoo." *Journal of Popular Culture* 37, no. 3 (February 2004): 379–98.

Benbow, S. Mary P., and Bonnie C. Hallman. "Reading the Zoo Map: Cultural Heritage Insights from Popular Cartography." *International Journal of Heritage Studies* 14, no. 1 (2008): 30–41.

Bender, Daniel E. *The Animal Game: Searching for Wildness in the American Zoo.* Cambridge, MA: Harvard University Press, 2016.

Bennett, Brett M., and Joseph M. Hodge, eds. *Science and Empire: Knowledge and Networks of Science across the British Empire, 1800–1970.* Houndmills, Basingstoke, Hampshire: Palgrave Macmillan, 2011.

Bennett, Malcolm. "Vets: Historically the Most Dangerous Animal in the Zoo . . . Now out in the Wild!" In *History of Zoos and Aquariums: From Royal Gifts to Biodiversity Conservation,* edited by G. M. Reid and G. Moore, 132–38. Chester: North of England Zoological Society, 2014.

Benson, Etienne. "Animal Writes: Historiography, Disciplinarity and the Animal Trace." In *Making Animal Meaning,* edited by Linda Kalof and Georgina M. Montgomery, 3–16. East Lansing: Michigan State University Press, 2011.

Berger, John. "Why Look at Animals?," 1–26. In *About Looking.* London: Readers and Writers, 1980.

Bieder, Robert E. *Bear.* London: Reaktion, 2005.

Birke, Linda. "Animal Bodies in the Production of Scientific Knowledge: Modelling and Medicine." *Body & Society* 18 (2012): 156–78.

———. "Intimate Familiarities? Feminism and Human-Animal Studies." *Society & Animals* 10, no. 4 (2008): 429–36.

Blunt, Wilfrid. *The Ark in the Park: The Zoo in the Nineteenth Century.* London: Book Club Associates, 1976.

Bostock, Stephen St. C. *Zoos and Animal Rights: The Ethics of Keeping Animals.* London: Routledge, 1993.

Bousé, Derek. *Wildlife Films.* University Park: University of Pennsylvania Press, 2000.

Boyd-Orr, John. *Food, Health and Income.* London: Macmillan, 1936.

Braitman, Laurel. *Animal Madness: How Anxious Dogs, Compulsive Parrots, and Elephants in Recovery Help Us Understand Ourselves.* New York: Simon & Schuster, 2014.

Branagan, D. F. "Stutchbury, Samuel (1798–1859)." *Oxford Dictionary of National Biography.* Oxford: Oxford University Press, 2004. http://www.oxforddnb.com/view/article/38030.

Brantz, Dorothee, ed. *Beastly Natures: Animals, Humans and the Study of History.* Charlottesville: University of Virginia Press, 2010.

Brassey, Charlotte A., et al. "Finite Element Modelling versus Classic Beam Theory: Comparing Methods for Stress Estimation in a Morphologically Diverse Sample of Vertebrate Long Bones." *Journal of the Royal Society: Interface* 10 (2012): 1–11.

Braverman, Irus. *Zooland: The Institution of Captivity.* Stanford: Stanford Law Books, 2013.

Brockway, Lucile H. *Science and Colonial Expansion: The Role of the British Royal Botanic Gardens.* London: Academic Press, 1979.

Broom, Donald M. "A History of Animal Welfare Science." *Acta Biotheor* 59 (2011): 121–37.

Brown, Tim, Alan Ashby, and Christoph Schwitzer. *An Illustrated History of Bristol Zoo Gardens.* Todmorden, West Yorkshire: Independent Zoo Enthusiasts Society, 2011.

Brunner, Bern. *The Ocean at Home: An Illustrated History of the Aquarium.* Translated by Ashley Marc Slapp. London: Reaktion, 2011.

Burns, Ann, "Examining the Hormonal and Behavioural Effects of Using Follicle Stimulating Hormone (FSH) Injection and Oral Clomiphene for Ovarian Stimulation in a Captive Western Lowland Gorilla." MSc diss., University of Bristol, 2014.

Burghardt, Gordon M., and Harold A. Herzog, Jr. "Animals, Evolution, and Ethics." In *Perceptions of Animals in American Culture,* edited by R. J. Hoage, 129–51. London: Smithsonian Institution Press, 1989.

Burt, Jonathan. *Animals in Film.* London: Reaktion, 2002.

———. "The Illumination of the Animal Kingdom: The Role of Light and Electricity in Animal Representation." *Society & Animals* 9, no. 3 (2001): 203–28.

Burton, Antoinette. *At the Heart of the Empire: Indians and the Colonial Encounter in Late Victorian Britain.* Berkeley: University of California Press, 1998.

Byers, Onnie, Caroline Lees, Jonathan Wilken, and Christoph Schwitzer. "The One Plan Approach: The Philosophy and Implementation of CBSG's Approach to Integrated Species Conservation Planning." *WAZA Magazine* 14 (2013), "Toward Integrated Species Conservation" (2013): 2–5.

Callon, Michel. "Some Elements of a Sociology of Translation: Domestication of the Scallops and the Fishermen of Saint Brieuc Bay." In *Power, Action and Belief: A New Sociology of Knowledge?,* edited by J. Law, 196–223. London: Routledge, 1986.

Candea, Matai. "'I Fell in Love with Carlos the Meerkat': Engagement and Detachment in Human-Animal Relations." *American Ethnologist* 37, no. 2 (2010): 241–58.

Carr, Neil, and Scott Cohen. "The Public Face of Zoos: Images of Entertainment, Education and Conservation." *Anthrozoös* 24, no. 2 (2011): 175–89.

Carruthers, D. "The Enlightenment Science of Society." In *Inventing Human Science: Eighteenth-Century Domains,* edited by Christopher Fox, Roy Porter, and Robert Wokler, 232–70. Berkeley: University of California Press, 1995.

Cartmill, Matt. *A View to a Death in the Morning: Hunting and Nature through History.* Cambridge, MA: Harvard University Press, 1993.

Cavalieri, Paola, and Peter Singer, eds. *The Great Ape Project: Equality beyond Humanity.* London: Fourth Estate, 1993.

Chakrabarty, Dipesh. "The Climate of History: Four Theses." *Critical Enquiry* 35 (2009): 197–222.

Chambers, Paul. *Jumbo: The Greatest Elephant in the World.* London: André Deutsch, 2007.

Clarke, A. G. "The Frozen Ark Project: The Role of Zoos and Aquariums in Preserving the Genetic Material of Threatened Animals." *International Zoo Yearbook* 43 (2009): 222–30.

Clarke, Ernest. "Spooner, Charles (1806–1871)." Revised by Linda Warden. *Oxford Dictionary of National Biography.* Oxford: Oxford University Press, 2004. http://www.oxforddnb.com/view/article/26161.

Classen, Constance, ed. *The Book of Touch.* Oxford: Berg, 2005.

Coates, Peter A. "Creatures Enshrined: Wild Animals as Bearers of Heritage." *Past and Present* 226, supplement 10 (2015).

———. *Nature: Western Attitudes since Ancient Times.* Cambridge: Polity, 1998.

———. "The Strange Stillness of the Past: Towards an Environmental History of Sound and Noise." *Environmental History* 10, no. 4 (2005): 636–65.

Colby, R. H., C. A. Mattacks, and C. M. Pond. "Gross Anatomy, Cellular Structure and Fatty Acid Composition of Adipose Tissue in Captive Polar Bears (Ursus maritimus)." *Zoo Biology* 12 (1993): 267–75.

Coleman, Jon. *Vicious: Wolves and Men in America.* New Haven: Yale University Press, 2004.

Collard, Rosemary-Claire, and Jessica Dempsey. "Life for Sale? The Politics of Lively Commodities." *Environment and Planning A* 25, no. 11 (2014): 2682–99.

Conway, William G. "Zoos: Their Changing Roles." *Science,* n.s., 163, no. 3862 (3 January 1969): 48–52.

Corbey, Raymond, and Ben Theunissen, eds. *Ape, Man, Apeman: Changing Views since 1600, Evaluative Proceedings of the Symposium "Ape, Man, Apeman: Changing Views since 1600, Leiden, the Netherlands 28 June-1 July 1993."* Leiden: Leiden University, 1995.

Corbin, Alain. *The Foul and the Fragrant: Odour and the French Social Imagination.* Cambridge, MA: Harvard University Press, 1986.

Cowie, Helen. "Elephants, Education and Entertainment: Travelling Menageries in Nineteenth-Century Britain." *Journal of the History of Collections* 25, no. 1 (2012): 1–19.

———. *Exhibiting Animals in Nineteenth-Century Britain: Empathy, Education, Entertainment.* London: Palgrave Macmillan, 2015.

Creager, A. N. H., and W. C. Jordan, eds. *The Human-Animal Boundary: Historical Perspectives.* Woodbridge, Suffolk: University of Rochester Press, 2002.

Crist, Eileen. *Images of Animals: Anthropomorphism and Animal Mind.* Philadelphia: Temple University Press, 1999.

Crosby, Alfred W. *Ecological Imperialism: The Biological Expansion of Europe, 900–1900.* Cambridge: Cambridge University Press, 1986.

Crutzen, P. J., and E. F. Stoermer. "The 'Anthropocene': An Epoch of Our Making." *Global Change Newsletter* 41 (2000): 17–18.

Cunningham, A. "Transforming Plague: The Laboratory and the Identity of Infectious Diseases." In *The Laboratory Revolution in Medicine,* edited by A. Cunningham and P. Williams, 209–44. Cambridge: Cambridge University Press, 1992.

Cunningham, H. *Leisure in the Industrial Revolution, c.1780–c.1880.* London: Croom Helm, 1980.

Cushing, N., and K. Markwell. "Platypus Diplomacy: Animal Gifts in International Relations." *Journal of Australian Studies* 33, no. 3 (2009): 255–71.

Darnton, Robert. *The Great Cat Massacre and Other Episodes in French Cultural History.* London: Allan Lane, 1984.

Daston, Lorraine, and Gregg Mitman, eds. *Thinking with Animals: New Perspectives on Anthropomorphism.* New York: Columbia University Press, 2005.

Davies, Gail. "Networks of Nature: Stories of Natural History Film-Making from the BBC." PhD diss., University College London, 1998.

Davis, Susan G. *Spectacular Nature: Corporate Culture and the Sea World Experience.* Berkeley: University of California Press, 1997.

de Courcy, Catherine. *Dublin Zoo: An Illustrated History.* Doughcloyne: Collins Press, 2009.

Derrida, Jacques. "The Animal That Therefore I Am (More to Follow)." *Critical Inquiry* 28, no. 2 (2002): 369–418.

Desmond, Jane C. *Staging Tourism: Bodies on Display from Waikiki to Sea World.* Chicago: University of Chicago Press, 1999.

Drayton, Richard. *Nature's Government: Science, Imperial Britain and the "Improvement" of the World.* New Haven: Yale University Press, 2000.

Driver, Felix, and David Gilbert, eds. *Imperial Cities: Landscape, Display and Identity.* Manchester: Manchester University Press, 1999.

Eben Kirksey, S., and Stefan Helmreich. "The Emergence of Multispecies Ethnography." *Cultural Anthropology* 25, no. 4 (2010): 545–76.

Edwards, John. *London Zoo from Old Photographs.* London: John Edwards, 1996.

Ellenberger, Henri J. "The Mental Hospital and the Zoological Garden." In *Animals and Man in Historical Perspective,* edited by J. and B. Klaits, 59–93. New York: Harper & Row, 1974.

Emel, Jody. "Are You Man Enough, Big and Bad Enough? Ecofeminism and Wolf Eradication in the USA." *Environment and Planning D: Society and Space* 13 (1995): 707–35.

Evans, E. P. *The Criminal Prosecution and Capital Punishment of Animals.* London: W. Heinemann, 1906.

Everest, Sophie. "'Under the Skin': The Biography of a Manchester Mandrill." In *The Afterlives of Animals: A Museum Menagerie,* edited by Samuel J. M. M. Alberti, 75–91. Charlottesville: University of Virginia Press, 2011.

Fan, Fa-ti. *British Naturalists in Qing China: Science, Empire and Cultural Encounter.* Cambridge, MA: Harvard University Press, 2004.

Fericean, Mihaela Liana, *et al.* "The History and Development of Ethology." *Research Journal of Agricultural Science* 47, no. 2 (2015): 45–51.

Fisher, James. *Zoos of the World: The Story of Animals in Captivity.* London: Aldus Books, 1966.

Flack, Andrew J. P. "Capturing the Beasts: Zoo Film and Interspecies Pasts." In *The Zoo and Screen Media: Images of Exhibition and Encounter,* edited by Michael Lawrence and Karen Lury, 23–42. Basingstoke: Palgrave Macmillan, 2016.

———. "'The Illustrious Stranger': Hippomania and the Nature of the Exotic." *Anthrozoös* 26, no. 1 (2013): 43–59.

———. "In Sight, Insane: Animal Agency, Captivity and the Frozen Wilderness in the Late Twentieth Century." *Environment and History* 22, no. 4 (2016): 629–52.

———. "Lions Loose on a Gentleman's Lawn: Animality, Authenticity and Automobility in the Emergence of the English Safari Park." *Journal of Historical Geography* 54 (October 2016): 38–49.

Foltz, Richard C. "Does Nature Have Historical Agency? World History, Environmental History, and How Historians Can Help Save the Planet." *History Teacher* 37, no. 1 (2003): 9–28.

Foucault, M. *Discipline and Punish: The Birth of the Prison.* Translated by Alan Sheridan. London: Allen Lane, 1977.

———. *Power/Knowledge: Selected Interviews and Other Writings.* Edited by Colin Gordon; translated by Colin Gordon et al. Brighton: Harvester, 1980.

Fowler, Murray E. "Historical Perspective of Zoo and Wildlife Medicine." *Journal of Veterinary Medicine Education* 33, no. 3 (2006): 326–30.

Franklin, Adrian. *Animals and Modern Cultures: A Sociology of Human-Animal Relations in Modernity.* London: Sage, 1999.

French, Richard D. *Antivivisection and Medical Science.* Princeton: Princeton University Press, 1975.

Frost, Warwick, ed. *Zoos and Tourism: Conservation, Education and Entertainment?* Bristol: Channel View, 2011.

Fudge, Erica. *Animal.* London: Reaktion, 2002.

———. "A Left-Handed Blow." In *Representing Animals,* edited by Nigel Rothfels, 3–18. Bloomington: Indiana University Press, 2002.

———. *Perceiving Animals: Humans and Beasts in Early Modern English Culture.* Basingstoke: Macmillan, 1999.

———. "Renaissance Animal Things." In *Gorgeous Beasts: Animal Bodies in Historical Perspective,* edited by Joan B. Landes, Paula Young Lee, and Paul Youngquist, 41–56. University Park: Pennsylvania State University Press, 2012.

Fuentes, Agustin. "Monkey and Human Interconnections: The Wild, the Captive, and the In-Between." In *Where the Wild Things Are Now: Domestication Reconsidered,* edited by Rebecca Cassidy and Molly H. Mullin, 123–45. Oxford: Berg, 2002.

Gardiner, A. "The 'Dangerous' Women of Animal Welfare: How British Veterinary Practice Went to the Dogs." *Social History of Medicine* 27, no. 3 (2014): 466–87.

Gott, Ted, and Kathryn Weir. *Gorilla.* London, Reaktion, 2013.

Green, Susie. *Tiger.* London: Reaktion, 2008.

Green-Armytage, A. H. N. *Bristol Zoo, 1835–1965: A Short History of the Bristol, Clifton and West of England Zoological Society.* Bristol: J. W. Arrowsmith, 1964.

Gregory, James. *Of Victorians and Vegetarians: The Vegetarian Movement in Nineteenth-Century Britain.* London: Tauris Academic Studies, 2007.

Gregory, W. K., and Henry C. Raven. *In Quest of Gorillas.* New Bedford: Darwin Press, 1937.

Grier, Katherine C. *Pets in America: A History.* Chapel Hill: University of North Carolina University Press, 2006.

Griffin, Donald R. *The Question of Animal Awareness: Evolutionary Continuity of Mental Experience.* New York: Rockefeller University Press, 1976.

Grove, Richard H. *Green Imperialism: Colonial Expansion, Tropical Island Edens and the Origins of Environmentalism, 1600–1960.* Cambridge: Cambridge University Press, 1995.

Gruffudd, Pyrs. "Biological Cultivation: Lubetkin's Modernism at London Zoo in the 1930s." In *Animal Spaces, Beastly Places: New Geographies of Human-Animal Relations,* edited by Chris Philo and Chris Wilbert, 2225–42. London: Routledge, 2000.

Hahn, Daniel. *The Tower Menagerie: The Amazing 600-Year History of the Royal Collection of Wild and Ferocious Beasts at the Tower of London.* New York: Jeremy P. Tarcher/Penguin, 2004.

Hallman, Bonnie C., and S. Mary P. Benbow. "Canadian Human Landscape Examples: Naturally Cultural: The Zoo as Cultural Landscape." *Canadian Geographer* 50, no. 2 (2006): 256–64.

Hancocks, David. *A Different Nature: The Paradoxical World of Zoos and Their Uncertain Future.* Berkeley: University of California Press, 2001.

Hanson, Elizabeth. *Animal Attractions: Nature on Display in American Zoos.* Princeton: Princeton University Press, 2002.

Haraway, Donna J. *Primate Visions: Gender, Race and Nature in the World of Modern Science.* New York: Routledge, 1989.

———. *When Species Meet.* Minneapolis: University Minnesota Press, 2008.

Hardy, Anne. "Professional Advantage and Public Health: British Veterinarians and State Veterinary Services, 1865–1939." *Twentieth Century British History* 14 (2003): 1–23.

Harfield, Jes. "Philosophical Ethology: On the Extents of What It Is to Be a Pig." *Society & Animals* 19, no. 1 (2011): 83–101.

Hart, Lynette A. "The Elephant-Mahout Relationship in India and Nepal: A Tourist Attraction." In *Animals in Person: Cultural Perspectives on Human-Animal Intimacy,* edited by John Knight, 163–89. Oxford: Berg, 2005.

Headrick, Daniel R. *The Tools of Empire: Technology and European Imperialism in the Nineteenth Century.* Oxford: Oxford University Press, 1981.

Hearne, Vicki. *Adam's Task: Calling Animals by Name.* London: Heinemann, 1987.

Henninger-Voss, Mary J., ed. *Animals in Human Histories: The Mirror of Nature and Culture.* Rochester, NY: University of Rochester Press, 2002.

Hoage, R. J., and William D. Deiss, eds. *New Worlds, New Animals: From Menagerie to Zoological Park in the Nineteenth Century.* Baltimore: Johns Hopkins University Press, 1996.

Hoage, R. J., Anne Roskell, and Jane Mansour. "Menageries and Zoos to 1900." In Hoage and Deiss, *New Worlds, New Animals,* 8–18.

Hoffer, Peter Charles. *Sensory Worlds in Early America.* Baltimore: Johns Hopkins University Press, 2003.

Hosey, Geoff. "Hediger Revisited: How Do Zoo Animals See Us?" *Journal of Applied Animal Welfare Science* 16, no. 4 (2013): 338–59.

Howell, Phillip. *At Home and Astray: The Domestic Dog in Victorian Britain.* Charlottesville: University of Virginia Press, 2015.

Hribal, Jason C. "Animals, Agency and Class: Writing the History of Animals from Below." *Human Ecology Review* 14, no. 1 (2007): 101–12.

———. *Fear of the Animal Planet: The Hidden History of Animal Resistance.* Oakland, CA: AK Press, 2010.

Hyson, Jeffrey Nugent. "Urban Jungles: Zoos and American Society." PhD diss., Cornell University, 1999.

Ingold, Tim. *Lines: A Brief History.* London: Routledge, 2007.

Inkster, Ian. "Introduction: Aspects of the History of Science and Science Culture in Britain, 1780–1850 and Beyond." In *Metropolis and Province: Science in British Culture, 1780–1850,* edited by Ian Inkster and Jack Morrell, 11–54. London: Hutchinson, 1983.

Ito, Takashi. *London Zoo and the Victorians, 1828–59.* Woolbridge, Suffolk: Boydell Press, 2014.

Itoh, Mayumi. *Japanese Wartime Zoo Policy: The Silent Victims of World War II.* New York: Palgrave Macmillan, 2010.

Jackson, Deirdre. *Lion.* London: Reaktion, 2010.

Jolly, W. P. *Jumbo.* London: Constable, 1976.

Jones, Donald. *A History of Clifton.* London: Philimore, 1992.

Jones, Robert W. "The Sight of Creatures Strange to Our Clime: London Zoo and the Consumption of the Exotic." *Journal of Victorian Culture* 2, no. 1 (1997): 1–26.

Jupp, Peter C., and Clare Gittings, eds. *Death in England: An Illustrated History.* Manchester: Manchester University Press, 1999.

Kalof, Linda. *Looking at Animals in Human History.* London: Reaktion, 2007.

Kalof, Linda, and Georgina M. Montgomery, eds. *Making Animal Meaning.* East Lansing: Michigan State University Press, 2011.

Kalof, Linda, and Brigitte Resl, gen. eds. *A Cultural History of Animals.* 6 vols. Oxford: Berg, 2007.

Kean, Hilda. *Animal Rights: Political and Social Change in Britain since 1800.* London: Reaktion, 1998.

———. "Challenges for Historians Writing Animal-Human History." *Anthrozoös* 25, supplement (2012): S57-S72.

———. *The Great Cat and Dog Massacre: The Real Story of World War Two's Unknown Tragedy.* Chicago: University of Chicago Press, 2017.

———. "Human and Animal Space in Historic 'Pet' Cemeteries in London, New York and Paris." In *Animal Death,* edited by J. Johnston and F. Probyn-Rapsey, 21–42. Sydney: Sydney University Press, 2013.

———. "The Moment of Greyfriars Bobby: The Changing Cultural Position of Animals, 1800–1921." In *A Cultural History of Animals in the Age of Empire,* edited by Kathleen Kete, 25–46. Oxford: Berg, 2007.

Keeling, C. H. *The Bristol Book.* Shelford: Clam, 1998.

———. *Where the Lion Trod: A Study of Forgotten Zoological Gardens.* N.p.: Clam, 1984.

Kelly, D. P., H. Pearson, and A. I. Wright. "Morbidity in Captive White Tigers."

In *The Comparative Pathology of Zoo Animals,* edited by Richard J. Montali and George Migaki, 183–88. Washington, DC: Smithsonian Institution Press, 1980.

Kerr, Adrian, and Lynn Steinberg, eds. *Mourning Diana: Nation, Culture, and the Performance of Grief.* London: Routledge, 1999.

Kete, Kathleen. *The Beast in the Boudoir: Pet-Keeping in Nineteenth-Century Paris.* Berkeley: University of California Press, 1994.

———, ed. *A Cultural History of Animals in the Age of Empire.* Oxford: Berg, 2007.

Kirksey, S. Eban, and Stefan Helmreich, "The Emergence of Multispecies Ethnography." *Cultural Anthropology* 25, no. 4 (2010): 545–76.

Kisling, Vernon J., Jr., ed. *Zoo and Aquarium History: Ancient Animal Collections to Zoological Gardens.* London: CRC Press, 2000.

Kitchener, A. C., and C. Groves. "New Insights into the Taxonomy of *Macaca pagensis* of the Mentawai Islands, Sumatra." *Mammalia* 66, no. 4 (2002): 533–42.

Knight, John, ed. *Animals in Person: Cultural Perspectives on Human-Animal Intimacy.* Oxford: Berg, 2005.

———. "Feeding Mr. Monkey: Cross-Species Food "Exchange" in Japanese Monkey Parks." In Knight, *Animals in Person,* 231–53.

———. Introduction to Knight, *Animals in Person,* 1–13.

Laidlaw, Zoë. *Colonial Connections, 1815–1845: Patronage, the Information Revolution and the Colonial Government.* Manchester: Manchester University Press, 2005.

Lambert, David, and Alan Lester, eds. *Colonial Lives across the British Empire: Imperial Careering in the Long Nineteenth Century.* Cambridge: Cambridge University Press, 2006.

Landes, Joan B., Paula Young Lee, and Paul Youngquist, eds. *Gorgeous Beasts: Animal Bodies in Historical Perspective.* University Park: Pennsylvania State University Press, 2012.

Lasdun, Susan. *The English Park: Royal, Private and Public.* London: Andre Deutsch, 1991.

Latour, B. *Reassembling the Social: An Introduction to Actor-Network-Theory.* Oxford: Oxford University Press, 2005.

Less, E. H, C. W. Kuhar, and K. E. Lukas. "Assessing the Prevalence and Characteristics of Hair-Plucking Behavior in Captive Western Lowland Gorillas (*Gorilla gorilla gorilla*)." *Animal Welfare* 22 (2013): 175–83.

Lever, Christopher. *They Dined on Eland.* London: Quiller Press, 1992.

Lévi-Strauss, Claude. *Totemism.* Translated by Rodney Needham. Uckfield, East Sussex: Beacon Press, 1963.

Lindahl Elliot, Nils. "See It, Sense It, Save It: Economies of Multisensuality in Contemporary Zoos." *Senses and Society* 1, no. 2 (2006), 203–24.

Lippit, Akira Mizuta. *Electric Animal: Towards a Rhetoric of Wildlife.* Minneapolis: University of Minnesota Press, 2000.

"Living Planet Report 2014." World Wide Fund for Nature, 2014. http://wwf
.panda.org/about_our_earth/all_publications/living_planet_report/.

Logan, Thad. *The Victorian Parlour: A Cultural Study.* Cambridge: Cambridge
University Press, 2001.

Lopez, B. H. *Of Wolves and Men.* New York: Charles Scribner's Sons, 1978.

Lorenz, Konrad. "Innate Forms of Potential Experience." *Z Tierpsychol* 5 (1943):
235–409.

Lorimer, Jamie. "Nonhuman Charisma: Which Species Trigger Our Emotions and
Why?" *ECOS* 27, no. 1 (2006): 20–27.

Lynn, Martin. "British Business and the African Trade: Richard & William King
Ltd. of Bristol and West Africa, 1833–1918." *Business History* 34, no. 4 (1992):
20–37.

———. "From Sail to Steam: The Impact of the Steamship Services on the British
Palm Oil Trade with West Africa, 1850–1890." *Journal of African History* 30
(1989): 227–45.

Macdonald, Sharon. "Expanding Museum Studies: An Introduction." In *A Companion to Museum Studies,* edited by Sharon Macdonald. Oxford: Blackwell,
2006.

MacKenzie, John M. *The Empire of Nature: Hunting, Conservation and British Imperialism.* Manchester: Manchester University Press, 1988.

———, ed. *Imperialism and Popular Culture.* Manchester: Manchester University
Press, 1986.

———, ed. *Imperialism and the Natural World.* Manchester: Manchester University Press, 1990.

Maddeaux, Sarah Joy. "'Our Clifton Zoo': A Social History of Bristol Zoo Gardens
since 1835." PhD diss., University of Bristol, 2014.

Malamud, Randy, ed. *A Cultural History of Animals in the Modern Age.* Oxford:
Berg, 2007.

———. *Reading Zoos: Representations of Animals and Captivity.* Basingstoke:
Macmillan, 1998.

Marsh, Jon. "The Rise of Celebrity Culture." In *Charles Dickens in Context,* edited
by Sally Ledger and Holly Furneaux, 98–108. Cambridge: Cambridge University Press, 2011.

McCann, J. *Green Land, Brown Land, Black Land: An Environmental History of
Africa.* Portsmouth, NH: Heinemann, 1999.

McCracken, Donal P. *Gardens of Empire: Botanical Institutions of the Victorian
British Empire.* Leicester: Leicester University Press, 1997.

McFarland, Sarah E., and Ryan Hediger, eds. *Animals and Agency: An Interdisciplinary Exploration.* Boston: Brill, 2009.

Mellen, Jill D., David J. Shepherdson, and Michael Hutchins. "Epilogue: The
Future of Environmental Enrichment." In *Second Nature: Environmental Enrichment for Captive Animals,* edited by David J. Sheperdson, Jill D. Mellen, and

Michael Hutchins, 329–36. Washington, DC: Smithsonian Institution Press, 1998.

Meller, Helen. *Leisure and the Changing City, 1870–1914.* London: Routledge and Kegan Paul, 1976.

Merchant, Carolyn. *The Death of Nature: Women, Ecology and the Scientific Revolution.* New York: Harper & Row, 1979.

Metcalf, Thomas R. *Imperial Connections: India in the Indian Ocean Arena, 1860–1920.* Berkeley: University of California Press, 2007.

Meyer, Fred A. "The New Animal Hospital and Quarantine Station at the London Zoo." *International Zoo Yearbook* 1, no. 1 (January 1960): 47–50.

Mikhail, Alan. *The Animal in Ottoman Egypt.* New York: Oxford, 2014.

Miller, David Phillip, and Peter Hans Reill, eds. *Visions of Empire: Voyages, Botany and Representations of Nature.* Cambridge: Cambridge University Press, 1996.

Miller, Ian Jared. *The Nature of the Beasts: Empire and Exhibition at the Tokyo Imperial Zoo.* Berkeley: University of California Press, 2013.

Mitchell, Timothy. *Colonising Egypt.* Berkeley: University of California Press, 1991.

Mitman, Gregg. "Ebola in a Stew of Fear." *New England Journal of Medicine,* 17 September 2014. http://www.nejm.org/doi/full/10.1056/NEJMp1411244#t=article.

———. "Pachyderm Personalities: The Media of Science, Politics and Conservation." In *Thinking with Animals: New Perspectives on Anthropomorphism,* edited by Lorraine Daston and Gregg Mitman, 175–95. New York: Columbia University Press: 2005.

———. *Reel Nature: America's Romance with Wildlife on Film.* Cambridge, MA: Harvard University Press, 1999.

———. "When Nature *Is* the Zoo: Vision and Power in the Art and Science of Natural History." *Osiris* 2, no. 11 (1996): 117–43.

Morris, Desmond. *The Naked Ape: A Zoologist's Study of the Human Animal.* London: Jonathan Cape, 1967.

Mullan, Bob, and Garry Marvin. *Zoo Culture.* Urbana: University of Illinois Press, 1999.

Murray, Narisara. "Lives of the Zoo: Charismatic Animals in the Social Worlds of the Zoological Society of London, 1850–1897." PhD diss., Indiana University, 2004.

Nagel, Thomas. "What Is It Like to Be a Bat?" In *Mortal Questions.* Cambridge: Cambridge University Press, 1979: 165–80.

Nance, Susan, ed. *The Historical Animal.* Syracuse: Syracuse University Press, 2015.

Nash, Linda. "The Agency of Nature or the Nature of Agency." *Environmental History* 10, no. 1 (2005): 67–69.

Nash, Steve. "Desperately Seeking Charisma: Improving the Status of Invertebrates." *Bioscience* 54, no. 6 (2004): 487–94.

Neve, Michael. "Science in a Commercial City: Bristol, 1820–60." *Metropolis and*

Province: Science in British Culture, 1780–1850, edited by Ian Inkster and Jack Morrell, 179–204. London: Hutchinson, 1983.

O'Brien, S. J., P. Joslin, G. L. Smith, R. Wolfe, N. Schaffer, E. Heath, J. Ott-Joslin, P. P. Rawal, K. K. Bhattacharjee, and J. S. Martenson. "Evidence for African Origins of Founders of the Asiatic Lion Species Survival Plan." *Zoo Biology* 6 (1987): 99–116.

Osborne, Michael A. *Nature, the Exotic and the Science of French Colonialism.* Bloomington: Indiana University Press, 1994.

Paddon, Hannah. "Biological Objects and 'Mascotism': The Life and Times of Alfred the Gorilla." In *The Afterlives of Animals: A Museum Menagerie,* edited by Samuel J. M. M. Alberti, 135–48. Charlottesville: University of Virginia Press, 2011.

Patton, Paul, "Language, Power and the Training of Horses." In *Zoontologies: The Question of the Animal,* edited by C. Wolfe, 83–100. Minneapolis: University of Minnesota Press, 2003.

Pearson, Chris. "Between Instinct and Intelligence: Harnessing Police Dog Agency in Early Twentieth-Century Paris." *Comparative Studies in Society and History* 58, no. 2 (2016): 463–90.

Pearson, H., and A. I. Wright. "Some Observations on the Rearing of an Okapi Calf." *International Zoo Yearbook* 8 (1968): 134–36.

Peterson del Mar, David. "'Our Animal Friends': Depictions of Animals in Reader's Digest during the 1950s." *Environmental History* 3, no. 1 (1998): 25–44.

Philo, Chris, and Chris Wilbert, eds. *Animal Spaces, Beastly Places: New Geographies of Human-Animal Relations.* London: Routledge, 2000.

Phineas, Charles. "Household Pets and Urban Alienation." *Journal of Social History* 7, no. 3 (1974): 338–43.

Picker, John M. *Victorian Soundscapes.* Oxford: Oxford University Press, 2003.

Plumb, Christopher. "'In Fact, One Cannot See It without Laughing': The Spectacle of the Kangaroo in London, 1770–1830." *Museum History Journal* 3, no. 1 (2010): 7–32.

Plumwood, Val. "Being Prey." 2000. Available at *Kurungabaa: A Journal of Literature, History and Ideas from the Sea* 4 (18 January 2011), http://valplumwood.com/2008/03/08/being-prey/.

———. *Feminism and the Mastery of Nature.* London: Routledge, 1993.

Poliquin, Rachel. *The Breathless Zoo: Taxidermy and the Cultures of Longing.* University Park: Pennsylvania State University Press, 2012.

Porter, Bernard. *The Absent-Minded Imperialists: Empire, Society, and Culture in Britain.* Oxford: Oxford University Press, 2004.

Pratt, Mary Louise. *Imperial Eyes: Travel Writing and Transculturation.* London: Routledge, 1992.

Pyenson, Lewis, and Susan Sheets Pyenson. *Servants of Nature: A History of Scientific Institutions, Enterprises, and Sensibilities.* New York: W. W. Norton, 1999.

Quinn, Stephen Christopher. *Windows on Nature: The Great Habitat Dioramas of the American Museum of Natural History.* New York: Harry N. Abrams in association with the American Museum of Natural History, 2006.

Regan, Tom. *The Case for Animal Rights.* Berkeley: University of California Press, 1983.

Reid, G. M., and G. Moore, eds. *History of Zoos and Aquariums: From Royal Gifts to Biodiversity Conservation.* Chester: North of England Zoological Society, 2014.

Ritvo, Harriet. "Among Animals." *Environment and History* 20 (2014): 491–98.

———. *The Animal Estate: The English and Other Creatures in the Victorian Age.* Cambridge, MA: Harvard University Press, 1987.

———. "Animal Planet." *Environmental History* 9, no. 2 (2004): 204–20.

———. "Beasts in the Jungle (or Wherever)." *Daedalus* 137, no. 2 (Spring 2008): 22–30.

———. "Calling the Wild." In *Gorgeous Beasts: Animal Bodies in Historical Perspective,* edited by Joan B. Landes, Paula Young Lee, and Paul Youngquist, 105–16. University Park: Pennsylvania State University Press, 2012.

———. "The Emergence of Modern Pet-Keeping." *Anthrozoös* 1, no. 3 (1987): 158–65.

———. "History and Animal Studies." *Society & Animals* 10, no. 4 (2008): 403–6.

———. "The Order of Nature: Constructing the Collections of Victorian Zoos." In *New Worlds, New Animals,* 43–50.

———. *The Platypus and the Mermaid and Other Figments of the Classifying Imagination.* Cambridge, MA: Harvard University Press, 1997.

Robbins, Louise. *Elephant Slaves and Pampered Parrots: Exotic Animals in Eighteenth-Century Paris.* Baltimore: Johns Hopkins University Press, 2002.

Root, Nina J. "Victorian England's Hippomania." *Natural History* 2 (1993): 34–39.

Rothfels, Nigel. "Catching Animals." In *Animals in Human Histories: The Mirror of Nature and Culture,* edited by Mary J. Henninger-Voss, 182–228. Rochester, NY: University of Rochester Press, 2002.

———. "How the Caged Bird Sings: Animals and Entertainment." In *A Cultural History of Animals in the Age of Empire,* edited by Kathleen Kete, 95–112. Oxford: Berg, 2007.

———. "Immersed with Animals." In *Representing Animals,* edited by Nigel Rothfels, 199–223. Bloomington: Indiana University Press, 2002.

———, ed. *Representing Animals.* Bloomington: Indiana University Press, 2002.

———. *Savages and Beasts: The Birth of the Modern Zoo.* Baltimore: Johns Hopkins University Press, 2002.

———. "Touching Animals: The Search for a Deeper Understanding of Animals." In *Beastly Natures: Animals, Humans and the Study of History,* edited by Dorothee Brantz, 38–58. Charlottesville: University of Virginia Press, 2010.

Roychoudhury, A. K., and K. S. Samkhala. "Inbreeding in White Tigers." *Proceedings of the Indian Academy of Science* 88 (1979): 311–23.

Rupke, Nicholas. *Vivisection in Historical Perspective.* London: Croom Helm, 1987.

Ryder, Richard. "Experiments on Animals." In *Animals, Men and Morals: An Enquiry into the Mal-Treatment of Non-Humans,* edited by Stanley Godlovich, Rosalind Godlovich, and John Harris, 41–82. London: Victor Gollanoz, 1971.

Sack, Robert David. *Human Territoriality: Its Theory and History.* Cambridge: Cambridge University Press, 1986.

Salazar Parreńas, Rheana "Juno." "Producing Affect: Transnational Volunteerism in a Malaysian Orang-utan Rehabilitation Center." *American Ethnologist* 39, no. 4 (2012): 673–87.

Sanders, Clinton R. "Actions Speak Louder than Words: Close Relationships between Humans and Nonhuman Animals." *Symbolic Interaction* 26, no. 3 (2003): 405–26.

Sayer, K. "Animal Machines: The Public Response to Intensification in Great Britain, c. 1960–c. 1973." *Agricultural History* 87, no. 4 (2013): 473–501.

Sax, Boria. "The Basilisk and the Rattlesnake, or a European Monster Comes to America." *Society & Animals* 2, no. 1 (1994): 3–15.

———. *The Mythical Zoo: An Encyclopedia of Animals in World Myth, Legend and Literature.* Oxford: ABC-CLIO, 2001.

Schafer, R. Murray. "Soundscapes and Earwitnesses." In *Hearing History: A Reader,* edited by Mark M. Smith, 1–9. Athens: University of Georgia Press, 2004.

Schaller, George B. *The Serengeti Lion: A Study of Predator-Prey Relations.* Chicago: University of Chicago Press, 1972.

Scott, Shelly, R. "The Racehorse as Protagonist: Agency, Independence and Improvisation." In McFarland and Hediger, *Animals and Agency,* 45–65. Boston: Brill, 2009.

Serpell, James. *In the Company of Animals: A Study of Human-Animal Relationships.* Cambridge: Cambridge University Press, 1986.

Shapland, Andrew. "Where Have All the Monkeys Gone? The Changing Nature of the Monkey Temple at Bristol Zoo." *Anthrozoös* 17, no. 3 (2004): 194–209.

Shapland, Andrew, and David Van Reybrouck. "Competing Natural and Historical Heritage: The Penguin Pool at London Zoo." *International Journal of Heritage Studies* 14, no. 1 (2008): 10–29.

Shaw, David Gary. "A Way with Animals." *History and Theory* 52, no. 4 (2013): 1–12.

Shepherdson, David J., Jill D. Mellen, and Michael Hutchins, eds. *Second Nature: Environmental Enrichment for Captive Animals.* Washington, DC: Smithsonian Institution Press, 1998.

Siegel, Jonah. *The Emergence of the Modern Museum: An Anthology of Nineteenth-Century Sources.* Oxford, Oxford University Press, 2008.

Singer, Peter. *Animal Liberation.* London: Cape, 1975.

———. (ed.). *In Defense of Animals: The Second Wave.* London: Blackwell, 2006.

Skibins, Jeffrey C, Robert B. Powell, and Jeffrey C. Hallo. "Charisma and Conservation: Charismatic Megafauna's Influence on Safari and Zoo Tourists' Pro-Conservation Behaviors." *Biodiversity Conservation* 22 (2013): 959–82.

Smith, Mark M., ed. *Hearing History: A Reader.* Athens: University of Georgia Press, 2004.

Sorabji, R. *Animal Minds and Human Morals: The Origin of the Western Debate.* Ithaca: Cornell University Press, 1993.

Sorenson, John. *Ape.* London: Reaktion, 2009.

Specht, Joshua. "Animal History after Its Triumph: Unexpected Animals, Evolutionary Approaches and the Animal Lens." *History Compass* 14, no. 7 (2016): 326–36.

Spencer, Frank. "Pithekos to Pithecanthropus: An Abbreviated Review of Changing Scientific Views of the Relationship of Anthropoid Apes to Homo." In *Ape, Man, Apeman: Changing Views since 1600; Evaluative Proceedings of the Symposium "Ape, Man, Apeman: Changing Views since 1600, Leiden, the Netherlands 28 June–1 July 1993,"* edited by Raymond Corbey and Ben Theunissen, 13–22. Leiden: Leiden University, 1995.

Storey, William K. "Big Cats and Imperialism: Lion and Tiger Hunting in Kenya and Northern India, 1898–1930." *Journal of World History* 2, no. 2 (1991): 135–73.

Stutesman, Drake. *Snake.* London: Reaktion, 2005.

Swart, Sandra. *Riding High: Horses, Humans, and History in South Africa.* Johannesburg: Wits University Press, 2010.

Thomas, Keith. *Man and the Natural World: Changing Attitudes in England, 1500–1800.* London: Allen Lane, 1983.

Thompson, Andrew S. *The Empire Strikes Back? The Impact of Imperialism on Britain from the Mid-Nineteenth Century.* New York: Pearson Longman, 2005.

Tidabock, Tracy. "Captive Breeding Management in Livingstone's Fruit Bats (*Pteropus livingstoni*): Assessing Resultant Dominance and Reproductive Success." MSc diss. University of Bristol, 2014.

Tipper, Becky. "'A Dog Who I Know Quite Well': Everyday Relationships between Children and Animals." *Children's Geographies* 9, no. 2 (2011): 145–65.

Tuan, Ti-Fu. "The Pleasures of Touch." In *The Book of Touch,* edited by Constance Classen, 74–79. Oxford: Berg, 2005.

———. *Space and Place: The Perspective of Experience.* London: Edward Arnold, 1977.

Turner, James C. *Reckoning with the Beast: Animals, Pain and Humanity in the Victorian Mind.* Baltimore: Johns Hopkins University Press, 1980.

Uddin, Lisa. *Zoo Renewal: White Flight and the Animal Ghetto.* Minneapolis: University of Minnesota Press, 2015.

Van Sittert, Lance, and Sandra Swart. "Canis familiaris: A Dog History of South Africa." *South African Historical Journal* 48, no. 1 (2003): 138–73.

Veltre, Thomas. "Menageries, Metaphors and Meanings." In *New Worlds, New Animals: From Menagerie to Zoological Park in the Nineteenth Century,* edited by R. J. Hoage and William A. Deiss, 19–29. Baltimore: Johns Hopkins University Press, 1996.

Vialles, Noilie. *Animal to Edible.* Translated by J. A. Underwood. Cambridge: Cambridge University Press, 1994.

Vickery, Amanda. "Golden Age to Separate Spheres? A Review of the Categories and Chronology of English Women's History." *Historical Journal* 36, no. 2 (1993): 383–414.

Wadewitz, Lissa. "Are Fish Wildlife?" *Environmental History* 16, no. 3 (2011): 423–27.

Walker, Brett L. "Animals and the Intimacy of History." *History and Theory* 52 (2013): 45–67.

Walvin, James. *A Child's World: A Social History of English Childhood, 1800–1914.* Hammondsworth: Penguin Books, 1982.

Wang, Ning. "Rethinking Authenticity in Tourism Experience." *Annals of Tourism Research* 26, no. 2 (1999): 349–70.

Warkentin, Traci. "Whale Agency: Affordances and Acts of Resistance in Captive Environments." In McFarland and Hediger, *Animals and Agency,* 23–43. Boston: Brill, 2009.

Warkentin, Traci, and Leesa Fawcett. "Whale and Human Agency in World-Making: Decolonizing Whale-Human Encounters." In *Metamorphoses of the Zoo: Animal Encounter after Noah,* edited by Ralph R. Acampora, 130–21. Lanham, MD: Lexington Books, 2010.

Warin, Robert, and Anne Warin. *Portrait of a Zoo: Bristol Zoological Gardens: 1835–1985.* Bristol: Redcliffe Press, 1985.

Webster, J. "University of Bristol School of Veterinary Science." *Veterinary Record* 141, no. 9, 30 August 1997.

Weeratunge, Chamalee. *The Elephant Gates.* Austin, TX: Greenleaf Book Group, 2014.

Weil, Kari. *Thinking Animals: Why Animal Studies Now?* New York: Colombia University Press, 2012.

Whatmore, S. *Hybrid Geographies: Natures, Cultures, Spaces.* London: Sage, 2002.

Willis-Tropea, Liz. "Hollywood Glamour: Sex, Power and Photography, 1925–39." PhD diss., University of Southern California, 2008.

Wilson, A. M. "Sexing the Hyena: Intraspecies Readings of the Female Phallus." *Signs* 28, no. 3 (2003): 762–70.

Wirtz, Patrick H. "Zoo City: Bourgeois Values and Scientific Culture in the Industrial Landscape." *Journal of Urban Design* 2, no. 1 (1997): 61–82.

Wolch, Jennifer, and Jody Emel, eds. *Animal Geographies: Places, Politics and Identity in the Nature-Culture Borderlands.* London: Verso, 1998.

———. "Guest Editorial: Bringing the Animals Back In." *Environment and Planning D: Society and Space* 13, no. 6 (1995): 632–38.

Wood, Amy Louise. "'Killing the Elephant': Murderous Beasts and the Thrill of Retribution, 1885–1930." *Journal of the Gilded Age and the Progressive Era* 11, no. 3 (2012): 405–44.

Woods, Abigail. "Rethinking the History of Modern Agriculture: British Pig Production, c. 1910–65." *Twentieth-Century British History* 23, no. 2 (2012): 165–91.

Woods, Abigail, and Stephen Matthews. "'Little, If at All, Removed from the Illit-
erate Furrier or Cow-Leech': The English Veterinary Surgeon, c. 1860–1885, and
the Campaign for Veterinary Reform." *Medical History* 54 (2010): 29–54.

Woods, Michael. "Fantastic Mr. Fox? Representing Animals in the Hunting
Debate." In *Animal Spaces, Beastly Places: New Geographies of Human-Animal
Relations,* edited by Chris Philo and Chris Wilbert, 182–202. London: Rout-
ledge, 2000.

Woodson-Boulton, Amy. "Victorian Museums and Victorian Society." *History
Compass* 6, no. 1 (2008): 109–146.

Worboys, Michael. "Was There a 'Bacteriological Revolution' in Late Nineteenth-
Century Medicine?" *Studies in History and Philosophy of Science Part C: Studies
in History and Philosophy of Biological and Biomedical Sciences* 38 (2007): 20–42.

Worster, Donald, *Nature's Economy: A History of Ecological Ideas.* San Francisco:
Sierra Club Books, 1977.

Wylie, Dan. *Elephant.* London: Reaktion, 2008.

Young, Robert M. *Darwin's Metaphor: Nature's Place in Victorian Culture.* Cambridge: Cambridge University Press, 1985.

Television

Rock, Chris, "Never Scared." *HBO.* April 2004.

Index

Italicized page numbers refer to illustrations, and BZ denotes Bristol Zoo.